WENDY TROY

SMP 11-16

Developing mathematical imagery

Activities for the classroom

The School Mathematics Project

Published by the Press Syndicate of the University of Cambridge
The Pitt Building, Trumpington Street, Cambridge CB2 1RP
40 West 20th Street, New York, NY 10011–4211, USA
10 Stamford Road, Oakleigh, Melbourne 3166, Australia

First published 1994

Printed in Great Britain by Bell and Bain Ltd., Glasgow

A catalogue record for this book is available from the British Library

ISBN 0 521 45977 X

Cover illustration by Linda Combi

WV

Contents

Preface [4]

Introduction [5]

Summary of activities [8]

The activities:

1 Starters [11]
2 Maths around us [15]
3 Plus and minus [19]
4 Imaginary cards [22]
5 Imaginary dice [24]
6 Boxes [28]
7 Bigger boxes [30]
8 Mental multiplication [33]
9 Matador [35]
10 Fives and threes [37]
11 Dominoes [39]
12 Posts and gaps [41]
13 Mind measures [44]
14 Footprints [48]
15 Patterns with numbers [50]
16 Remembering numbers [52]
17 Imaginary patterns [54]
18 Odds and evens [57]
19 Next please [59]
20 Tiles [64]
21 Chairs and squares [66]
22 Watch the angle [70]
23 Clock polygons [74]
24 Roundabout [80]
25 Everyday objects [83]
26 Back-to-back [85]
27 Points of view [89]
28 Visualising shapes [91]
29 Patterned polyhedra [95]
30 Shadows [99]
31 Folding paper [101]
32 Turnaround [105]
33 Elastic shapes [109]
34 Walkabout [111]
35 Worms [113]
36 Rings and strings [117]
37 Square dance [120]
38 Sorting out Eth Porter [123]
39 Words and their meanings [126]
40 Pictures with errors [127]

Bibliography [128]

Preface

The aim of this book is to provide activities which develop mathematical imagery and mathematical strategies. The activities vary in approach; for example some involve visualisation, some require pupils to interpret or construct diagrams, and some develop strategies for mental arithmetic. Throughout, the emphasis is on tasks which will develop and improve pupils' mental skills.

Each activity is a useful task in its own right, but will be more effective if incorporated into the programme of study for the class, either supplementing and enhancing existing curriculum materials or replacing some parts of them. The grid on pages 8–9 provides a summary of the activities and their main content, and should assist with planning.

The following people contributed to the writing of this book:

Tony Burghall	Peter Critchley
Michael Darby	Tony Gardiner
Bob Hartman	Diana Sharvill

In addition, thanks go to Alan Bishop for his suggestions, to Elaine Morgan for the artwork of the Chichester trail, to teachers in the schools that trialled the draft versions of the tasks, and to the editor at Cambridge University Press who was closely involved in their development.

Introduction

'Mathematics is a mental universe.' Hans Freudenthal

This book consists of a range of classroom activities designed to develop pupils' use of mathematical imagery. The aim is to create mental images of shape, movement, pattern and number, so that these images can be used in exploring mathematics. Pupils are encouraged to build their own images, but certain strategies that can be widely used are also introduced.

What is mental mathematics?

We took Freudenthal's statement "Mathematics is a mental universe" to mean:

- that *learning* mathematics is largely a matter of
  (a) creating a suitable mental universe of one's own, and
  (b) mastering the art of manipulating these mental images in ways which are consistent with the internal structure of that universe;
- that *doing* mathematics involves us in the activity of first creating or identifying the mental objects appropriate to a particular problem, and then manipulating these entities to solve the original problem.

Thus *mental mathematics* is *all mathematics*.

Why is mental mathematics important?

The idea that mental mathematics is all mathematics is rather well disguised in the typical syllabus or scheme of work. However, when considering the stages pupils need to go through to learn any particular syllabus item, we have to ask ourselves what mental processes are needed to support the mathematical thinking which that topic involves.

As an example, for most children numbers begin life attached to specific *objects* in the real world. Before very long, pupils perceive that the numbers themselves have an independent existence. The arithmetical operations are initially rooted in actual, or imagined, transformations of real objects. Addition involves the idea of combining two sets; subtraction stems from the real or imagined act of removing a subset; multiplication and division stem from counting so many equal groups, or sharing into so many equal groups. But this is only a beginning. Once the numbers which have to be combined get larger than about 20, children can no longer calculate reliably using concrete objects. Instead they have to develop mental images of numbers (for example, using place value and the usual representation in base 10), and reliable mental ways of manipulating these mental objects (for example, collecting units and tens separately before taking account of carries if these are needed).

It is often argued that pupils should be given a diagram to illustrate the situation described in a question. However, requiring pupils to draw their own diagrams is often the easiest, and most helpful, way of getting them to focus on the relevant mental images, which will help them to understand and solve the problem. Depriving pupils of this opportunity can leave them with a beautiful ready-made diagram but no way of thinking about it.

How will this book help pupils' mathematics?

The activities in this book are not designed merely to satisfy those National Curriculum statements which refer specifically to mental mathematics. Traditional 'mental arithmetic' allows those pupils who already have efficient mental strategies for doing arithmetic to practise and improve their strategies. This is a valuable activity, but all the teacher sees are the outcomes, not the processes. Moreover, once one begins to think about processes used in mental arithmetic, it soon becomes clear that they are also needed in many other areas of mathematics and have to be taught.

Our goal in writing this material has been to help teachers develop their pupils' mental strategies, rather than just practise those strategies which pupils happen to have developed themselves.

How can we help pupils develop mental strategies?

We give two rather different examples to illustrate the way in which this book can help pupils develop important mental strategies.

First example: One theme of the number-related activities in this book is the notion of mental 'boxes' – boxes in which we store raw and processed information; information which can be 'up-dated' (that is, replaced or transformed and then re-stored), with the current contents remaining available for retrieval when needed.

Much mathematical thinking can be analysed in terms of working with a pair of mental boxes (rather like a calculator with two memory cells and a display): we have to store some information in one box while processing that held in the other box. Whenever this process is needed, it is crucial that the contents of the first box are remembered accurately, and that they are not 'forgotten' or changed as a result of thinking about, or processing, the contents of the second box. *Such mental gymnastics can be taught and learned.*

Some pupils pick up these ideas for themselves but many others do not. Those who appear to succeed could often do much better if they were made aware of, and made to practise, the art of storing information in a retrievable form while processing other information. Without this skill it is hard to handle mental addition of two-digit numbers reliably, to say nothing of multiplying two-digit numbers or solving simple word problems involving more than one step.

Second example: Suppose a group has been challenged to fold a piece of paper in such a way that one straight cut will produce a triangular hole, or a pentagonal hole. Often they will be left to try to achieve this simply by trial and error. One consequence – apart from the huge waste of paper – is that pupils who 'succeed' may have no idea what it was they did in order to 'succeed', and so can neither explain nor repeat their 'success'.

A better approach is to encourage pupils, as soon as possible after their initial exploration of folding and cutting, to fold the paper in what they think might be a good way, but then to think, conjecture, discuss and decide what they expect the outcome of the intended cut to be, and why, before using the scissors. Such predictions force them to manipulate mental images of the folded paper, and eventually perhaps to discover for themselves the simple (but not obvious) strategy of imagining what shape the piece they plan to cut off will be when it is unfolded. Moreover, once pupils have committed themselves to reasoned predictions, rather than wild guesses, they are more likely to learn from the outcome when they eventually implement the final cut.

These two examples underline the fact that we are not advocating activity for its own sake. We want pupils to use and develop their imagination so that they can think about likely outcomes before acting or calculating, and afterwards reflect on and improve the images they have used.

Beyond a private mental universe to a shared universal mathematics

The activities suggested here encourage pupils to develop their own strategies, and to reflect on those strategies. However, developing one's own strategies is only the first stage of a much longer process.

It is certainly good practice to allow pupils who need to refer to concrete materials to do so. However, beyond some point, concrete materials tend to restrict rather than help pupils' thinking in mathematics. Thus our long-term aim must be to help pupils develop the ability to imagine suitable mental entities and to manipulate them effectively in the mind.

The second aspect we should mention concerns standard procedures, which occur in many parts of mathematics. These have usually become standard for good reasons. They are rarely the 'simplest' or most obvious methods, so one should not expect beginners to reinvent them spontaneously.

However, standard procedures are usually more efficient than primitive methods: they may generalise in ways that other methods do not, or they may simply provide a proven, flexible, common language which allows pupils, teachers and others to communicate meaningfully with each other. Thus pupils' own strategies should be seen as a means to an end, and not as an end in themselves. Once they have learned the necessary lessons from developing and reflecting on their own private strategies, they should be helped to move on to master any relevant standard procedures, so that these too become part of their own mental universe of mathematics.

Summary of the activities

This book contains 40 activities which cover a broad range of mathematical content, not just arithmetic. Each activity has been referenced in a grid on pages 8–9 to the most relevant strand in the National Curriculum. No level has been specified because we are focusing on the processes by which pupils tackle mental mathematics. In most cases the level at which they will be working will be higher than that to which the activity could be referenced.

The time each activity takes will depend upon many factors. The grid includes a rough indication of the time needed in one of three categories:

A – short activities, taking less than a lesson, which may be returned to on several occasions;
B – activities which are roughly one lesson long;
C – activities which are likely to take longer than a lesson.

Summary of activities

	Title	Brief description of activity	Main content strand in the National Curriculum	Time
1	Starters	These are brief activities which make pupils aware of their mental images.	General	A
2	Maths around us	The aim of this activity is to raise pupils' awareness of the mathematics that surrounds them in everyday life.	General	C
3	Plus and minus	This activity explores various strategies for adding and subtracting mentally.	Number: calculation	B
4	Imaginary cards	Calculations are done using imaginary number cards.	Number: calculation	B
5	Imaginary dice	This is the first of a series of activities designed to help pupils do mental calculations. It uses the visualisation of two dice to give pupils experience of storing integers mentally.	Number: calculation	B
6	Boxes	Pupils add simple numbers held in mental stores. This should follow 'Imaginary dice'.	Number: calculation	B
7	Bigger boxes	Pupils add two-digit numbers held in mental stores. This develops the ideas in 'Boxes'.	Number: calculation	B
8	Mental multiplication	Pupils use mental stores to help multiply a two-digit number by a one-digit number, using the skills developed in 'Bigger boxes'.	Number: calculation	B
9	Matador	This domino game encourages familiarisation with number bonds up to ten.	Number: calculation	A
10	Fives and threes	This domino game develops familiarity with multiples of three and five.	Number: calculation	A
11	Dominoes	These short activities use dominoes to develop mental arithmetic.	Number: calculation	A
12	Posts and gaps	Visual imagery is used to help with word problems where the answer is one more or one less than expected.	Number: calculation	B
13	Mind measures	The aim of this activity is to develop strategies for estimating lengths and areas.	Number: estimation	B
14	Footprints	Based on the evidence of a footprint, pupils have to find out as much as they can about the creature that made it.	Number: estimation, measurement, ratio	B/C
15	Patterns with numbers	This activity explores how it is easier to recognise how many objects there are if they are arranged in a pattern.	Number: estimation Algebra: number properties	B
16	Remembering numbers	Pupils hold a small set of numbers in their heads.	Algebra: number properties Number: calculation	A
17	Imaginary patterns	Pupils visualise the patterns in number grids and then answer questions about them.	Algebra: patterns and sequences	B
18	Odds and evens	This is the first of a set of activities which develops the understanding of algebraic expressions. It involves generalising simple sequences.	Algebra: patterns and sequences	B

19	Next please	This involves describing and generalising sequences based on geometrical patterns, and extends the work in 'Odds and evens'.	Algebra: patterns and sequences, expressions	B
20	Tiles	This follows 'Next please' and encourages pupils to visualise patterns of tiles in order to find algebraic rules.	Algebra: patterns and sequences, expressions	B
21	Chairs and squares	Pupils are encouraged to explain why apparently different algebraic expressions can be equivalent.	Algebra: expressions	B
22	Watch the angle	A clockface is used to help pupils visualise angles of a certain size.	Space: measures	B
23	Clock polygons	This activity uses a clockface to construct polygons mentally.	Space: shape	B+A
24	Roundabout	Pupils order a set of views of an object.	Space: shape	A/B
25	Everyday objects	Pupils identify or describe objects using mathematical terms.	Space: shape	B
26	Back-to-back	Pupils describe or draw simple diagrams using precise mathematical language.	Space: shape Using and applying: communication	B
27	Points of view	This activity develops pupils' ability to visualise and draw different views of an object.	Space: shape	B
28	Visualising shapes	A shape is visualised, then analysed or moved.	Space: shape	A
29	Patterned polyhedra	Pupils colour nets so that when they are made up, edges which touch have the same colour.	Space: shape	B
30	Shadows	Pupils imagine the shapes which are possible as shadows of various objects.	Space: shape	B
31	Folding paper	Pupils try to obtain a given shape by cutting a folded piece of paper.	Space: shape, movement	B
32	Turnaround	A triangle is fixed on a board which is then rotated. Pupils visualise the final position of the triangle.	Space: movement	B
33	Elastic shapes	Pupils visualise the new shapes made by moving vertices of certain two-dimensional shapes.	Space: movement	B
34	Walkabout	Pupils visualise and describe routes around their school.	Space: location	B
35	Worms	Pupils use multilink to construct shapes connecting the two ends of three-dimensional shapes.	Space: location	B
36	Rings and strings	Pupils visualise moving rings or strings to decide whether they are connected.	Space	B
37	Square dance	Pupils are asked to imagine a square with numbers in the corners and the results of swapping around some of these numbers.	Space	B
38	Sorting out Eth Porter	'Eth porter' is an anagram of 'the report'. Pupils put a set of phrases in a sensible order.	Using and applying: communication	B
39	Words and their meanings	This poster showing words drawn in a way to illustrate their meaning is used as a basis for discussion.	General	A/B
40	Pictures with errors	This poster contains a number of errors designed to encourage discussion.	General	A/B

The choice of activity will depend upon the experience and ability of the group. Many of the activities can be used with a whole class, although some of them may be more suitable for a small group or an individual. A small group should be able to tackle many of the activities with the teacher 'visiting' the group on a number of occasions throughout the lesson, leaving the teacher free to deal with the rest of the class for most of the time.

The format of the activities

The activities in this book are presented in a standard format. Each activity starts with up to four sub-headings.

Introduction A brief summary of the activity.
Materials A list of any materials required.
Possible content A list of the areas of mathematics likely to be encountered.
Preamble General notes regarding the use of the activity.

In general the subsequent text is split.

"Text on the left-hand side of each page indicates what the pupils are expected to do, or questions they may be asked. Quotes have been used for teacher questions, though it is **not intended that these should be used verbatim.**"

Text on the right-hand side of each page is a commentary on the more obvious or important aspects of what may happen. This part also includes ideas for the teacher, ways of working, and comments on the mathematics being developed.

At the end of a few activities there are possible extensions. These are not intended as a list to be worked through, but could be used as a source of activity at a deeper level for the more able pupil. They could also be a source of activity for anyone motivated by the original idea.

Preparing to use the activities

The style of working in the activities may be new to many teachers. The 'Starters' section includes a number of activities which could be used to familiarise both teachers and pupils with this type of activity.

Before using an activity with pupils, you must be familiar with the ideas involved in that activity. You should:

- read through the activity to get an overview;
- work through the task yourself, preferably with a colleague;
- make notes of any difficulties you may encounter;
- prepare any necessary resources.

1 Starters

Everyone has mental images and strategies which they use when doing mathematics. Because these are often very individualistic, people are reluctant to voice their thoughts in case they do not use a method which is perceived as the *correct* one.

These starter activities are designed to make pupils more aware of their mental images, and to show that these are acceptable and valuable. As was stated in the introduction, some images are more efficient than others, and this aspect will be explored in subsequent activities. However, at this initial stage it is important not to attach value judgements to pupils' ideas – with the possible exception of those which are actually wrong. The starters will also allow everyone to experience a diversity of images and from this realise that some can be more helpful than others.

The aptitude of the class will govern how long the tasks will take. They are intended to be fairly short, in order to introduce pupils to this novel way of thinking.

Careful handling is needed in the early stages. If the class responds well, then the ideas can be continued or extended. On the other hand, some pupils may find the situation 'uncomfortable' and will be reticent. Respect this, but as there is a very real benefit in getting past this mental barrier, encourage them to contribute. Avoid the temptation to stop too soon.

The activities can be done in any order.

■ 1 Give me five

"I want you to close your eyes."

This is to enable pupils to concentrate on what is in their heads, rather than what is around them.

"I don't want you to say anything. Just picture in your mind what I say. What do you see when I mention 'five'?"

Allow a short time for pupils to get an image.

"I expect that between you there are many different pictures. What you see is your picture, but it may not be the same as everyone else's. I would like us to share them. Do not be afraid to tell us what you saw; all pictures can be valuable."

"Tell me what you saw."

Collect responses on the board.
Discuss the images, but avoid value judgements.
Some likely responses are:

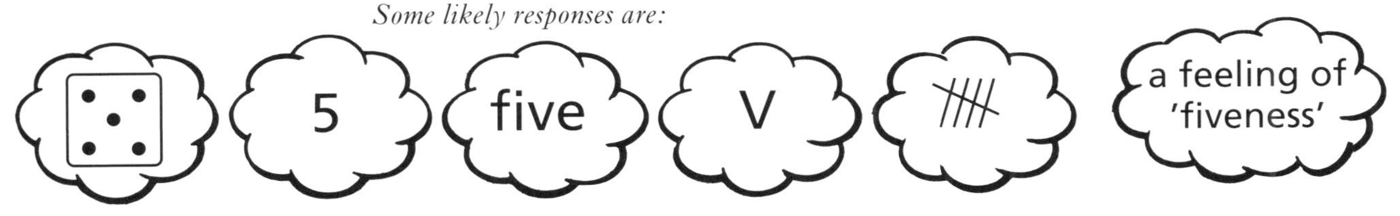

You may be surprised by the number of images generated. If appropriate, you could try another number, such as 16, to see if different images occur.

■ 2 Numbers everywhere

Make sure that each pupil has a pencil and plain paper.

"Close your eyes. I am going to give you a list of numbers. Don't concentrate on individual numbers, but try to picture the whole list."

Wait for silence. Watch the pupils' faces to judge how quickly to give the next number.

"One ... two ... three ... five ... nine ... ten ... fifteen ... twenty ... sixty ... one hundred ... three hundred ... two thousand ... a million ... four million ..."

There is nothing special about either the numbers in the list or the length of it, so choose numbers to suit the class.

Once you have completed your list, wait for pupils' images to form.

"Concentrate on what you saw. Open your eyes. Draw what you saw. Your picture may be different from other people's. This is quite acceptable."

After a suitable time, ask pupils to share what they have drawn.

Discuss as appropriate. Take care not to cause embarrassment as mental images are very personal.

■ 3 Banana panorama

"Close your eyes. We are going to do some mathematics in our minds. There is nothing to write down, but try to remember any thoughts you have."

You should pause between statements in the instructions to allow pupils to carry them out.

"Imagine a banana. Move it to the left. Move it up. Move it back to where it was to start with. Stand it up on one end. Spin it. Stop it spinning. Peel it. Throw the skin away. Split it in half. Turn one half one way, and the other half the other way. Put the two pieces back together again. Eat it."

Use whatever instructions you like, to suit the class.

"Open your eyes. What did you experience? What thoughts did you have? Did you find some things difficult to do? Which ones? Why?"

Discuss pupils' thoughts. Ask other questions as appropriate.

4 Nice and nasty

Make sure that pupils have pencil and paper.

"Think of a nice number. Write it down."

Pause.

"Think of a nasty number. Write it down."

Pause.

Ask pupils to say which numbers were nice/nasty and why. Discuss their choices and reasons. This activity should be continued by using other adjectives, such as friendly, peculiar, very big, very small, green, funny, and so on.

Discuss choices as appropriate.

5 Cube route

"Imagine a framework cube, with just rods along the edges."

Make sure that all pupils have a clear image of this. If you wish, you could ask some simple questions about it, such as: "Can you see the edges in groups of four?"

When giving the following instructions, leave pauses to allow pupils to imagine what is happening.

"Close your eyes. Move your cube so that the bottom is horizontal, just as if it had been put on a table. Put a mark on one of the bottom corners. Fred the fly lives at this corner. He walks along one of the bottom edges, and at the corner he carries on along another bottom edge. At the next corner he walks up an edge to a top corner, and along one of the top edges. He then remembers that he has left his sandwiches at home.

How would he get back home as quickly as possible by walking along the edges? Imagine his route. How many edges would he have to walk along?"

Allow time for pupils to picture this.

"Now open your eyes."

Discuss the answers to the questions, and get pupils to justify their answers. Avoid drawing anything, except as a last resort.

Continue the activity as appropriate by making up new routes. It is helpful to write down your routes in advance so that you can repeat them if necessary.

To make the situation more interesting, you could put Spike the spider at one of the vertices.

■ 6 Number Kim

You will need to display a set of numbers, remove them from view, and later show them again. This can be done by use of an OHP, a roller board or a large piece of paper.

Make sure that pupils have pencil and paper.

Write the numbers 7, 2, 12, 3, 4, 4, 8, 10, 5 and 6 in a random way like this:

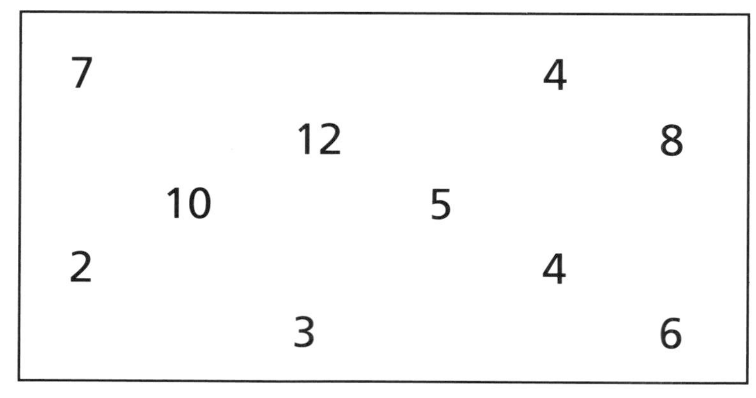

"Look at these numbers. I want you to try to remember them."

Leave the numbers on display for a couple of minutes – longer if you think it is necessary – and then cover them up.

"I want you to write down as many of those numbers as you can remember."

After a suitable time tell pupils to put their pencils down.
Ask them to tell you what the numbers were.
Ask other pupils if they agree or disagree.
Deal with missing or additional numbers as appropriate.

Discuss how they remembered the numbers and why missing numbers may have been left out.

Draw out that looking for patterns or sequences of numbers is a useful strategy. This can be repeated with other sets of numbers.
To emphasise the way that sequences help, your list could be composed of a single sequence, or two sequences. Or you could try putting the numbers in order …

Variations could be:

(a) to prepare a second display with some numbers on, and ask pupils if the two sets of numbers are the same. If they are not the same, ask what the changes were.
(b) to use shapes instead of numbers.

2 Maths around us

Introduction Most pupils do not realise how much they are surrounded by examples of mathematics. The aim of this activity is to raise their awareness.

Possible content Virtually anything.

Preamble The activity is broken down into stages so that you can decide where to stop. Some of the material may have to be adapted to suit your local circumstances. This will need careful planning. This activity could lead to the construction of a local maths trail.

"We are surrounded by mathematical symbols, ideas and objects. I want you to look around the classroom for items connected with mathematics. We are not just looking for numbers. Tell me what you can see."

Collect pupils' responses and write them on the board. Pupils are normally very imaginative and will suggest a wide range of items. If they do not, then give them leads such as 'time' or 'shape'. Discuss with the class anything which is unclear. The list can get surprisingly long.

"I want you to imagine that you are walking around the school. You should be able to 'see' other items which we do not have on our list. Tell me some."

Once again, collect and discuss. You may wish to ask pupils where they could see the items they mention.

Some pupils may find it difficult to change from the 'concrete' of the classroom to the 'abstract' of their imaginary walk, and may need help and encouragement.

"Now we are going to look even further afield. Imagine your journey home from school. What new mathematical items are there now?"

Moving out of the confines of the school should yield many more items from looking at shops, timetables, street furniture, etc.

Once again, collect and discuss.

As pupils will have a variety of routes home, you may wish to ask them to say where they would find the items they suggest.

Pupils living very close to the school may be asked to imagine a journey to, for example, the town centre.

The following is a useful homework task.

"We have done a lot of work in our minds. On the way home tonight, I want you to make a list of as many examples of mathematical items as you can, and also to write down where each one is."

Again, for those living close to the school you may wish to suggest a different route.

The list of locations is important if you are going to create a maths trail.

Next lesson you should encourage pupils to share ideas. For pupils who travel the same route it can be useful for them to compare their lists.

By this stage you should have a comprehensive list of 'maths around us'. Below we suggest three ways of using the information. You may think of others.

■ 1 A wall display

Pick out some of the more unusual items. Encourage pupils to write about them. You should try to include some sketches or photographs – just a list of items on the wall is not very exciting. The display can be as simple or as complex as suits the class.

■ 2 A scavenger hunt

Produce some sheets listing items and, for each item, ask the pupils to fill in exactly what they found and where they found it. This should be a practical exercise and not a mental one. If it is to be done in a restricted area, such as around the school, then try to ensure that all the items on your list can be found. A suggested format for a scavenger hunt sheet is shown here. The list below gives some ideas, but should be supplemented with ideas from your pupils.

BISHOP LUFFA SCHOOL
MATHS SCAVENGER HUNT

For each item on this list, say clearly what you found and where you found it.

Name: ....

Item	What	Where
a fraction *a sphere* *a date*		

Possible items for a scavenger hunt

A fraction; a decimal (not money); a percentage; a ratio; a prime number; a square number; a triangle number; a number in Roman numerals; a number larger than a million; something about 25m long; something about 10m high; something with a date on it; a time using a.m. or p.m.; a time using the 24-hour clock; a shape with an area between 1 and 2 square metres; a scale; a container holding between 300 ml and 500 ml; an object with a volume as near to 1 cubic metre as possible; an object weighing about 4kg; a formula; a number pattern; a square; a regular hexagon; a circle; a kite; a trapezium; a shape with rotational symmetry; a shape with exactly two lines of symmetry; a tessellation; a sphere; a cylinder; a prism (other than a cylinder); an acute angle; parallel lines; a spiral; an average; a line graph; a bar chart; a pie chart; an example of probability being used; something showing N, S, E or W.

3 A maths trail

This is the most time-consuming of the three suggestions, but can be the most rewarding.

You will need to decide on the area which the trail is going to cover (for example the school, the village, the town centre) and provide a large-scale map of the area. From the list of items, make a selection of a variety of objects and mark them on the map. Mapping pins are suitable for this. It can also be helpful to have a card index of items and locations so that various routes can be examined easily.

The compilation of your trail may be done by teachers, a small group of pupils, or separate groups dealing with different parts of the trail. However it is done, most of the planning should be done in school.

Once the trail is complete, try it out on colleagues or pupils who have not been involved in producing it. This will highlight any parts of the trail that are unclear, or directions which can be misunderstood. Modify the trail in the light of this and then get other classes to use it.

The first page of a trail is reproduced on page 18 as an example.

THE CHICHESTER TRAIL

This trail will take you through the streets of Chichester. There is a street map at the end of the booklet. None of the items mentioned is hidden, but you will need to be alert to make sure that you do not miss them. The trail should take you about $1^1/_2$ hours.

There are spaces for your answers. Any calculations or drawings can be done on the backs of the sheets.

x x x x x x x x x x x x x x x x x x x

Start outside the West Door of the Cathedral.

Estimate the height of the glass doors. ______________

Use this to estimate the height of the two towers. ______________

Behind you is the entrance to the Gardens of Prebendal School.
On one of the gateposts is a plaque. Read this.

When were the gardens opened? ______________

When was the school refounded? ______________ ←①

Walk towards the separate Bell Tower, which contains the Cathedral Shop.

②→ How long is the shop open each day? ______________

Nearby is a tree dedicated to Eline Morgan.
③→ How old was she when she died? ______________

Turn right, and walk towards the cross. You will pass the Post Office.

What is the time of the last collection on a Friday? ______________

What is this in the 24 hour clock? ______________

On the ground you will see a rectangular cover divided into smaller squares.
④→ How many squares are there? (Not 35!) ______________

Just nearby is another cover made up of six triangles. Sketch this.

One triangle would be 1/6 of the area.
What other fractions could you make? ______________

How many arches does the cross have? ______________

What is the name of a shape
with this many sides? ______________

Turn to face the way you have just come. You are looking West along West Street. What direction is the street on your left? ______________

Turn left and walk down this street. What is the number of the Offices of the Chichester Observer? ______________ ←⑤

There is an advert for photos in the window. Is the largest size print an exact enlargement of the smallest? ______________

Justify your answer.

____

____

① The plaque commemorates the 400th anniversary – so subtraction is necessary.

② Calculations are needed for this.

③ Eline's dates of birth and death are recorded.

④ This may have to be answered later.

⑤ The offices are not numbered. Adjacent properties will have to be looked at.

3 Plus and minus

Introduction The aim of this activity is to develop facility with mental addition and subtraction.

Materials Calculators for the games; possibly copies of sheet MI 1.

Possible content Mental addition and subtraction of numbers.

Preamble There is an initial teacher-led activity on addition, a second activity on subtraction, and finally two games which can be used to practise adding and subtracting mentally. The preliminary discussion aims to develop a variety of strategies. For part of the activity the pupils should work in pairs.

Traditionally, mental arithmetic has been a timed exercise. Here we are concentrating on strategies, so pupils should be given as much time as they need. Some pupils may not need to do addition and can move immediately to the subtraction activity.

"In your head, work out 59 + 37."

Write 59 + 37 on the board so pupils do not have a problem remembering the numbers. Give enough time for everyone to finish.

"How did you work it out?"

Collect all strategies on the board. Do not make any subjective comments. There are many equally valid ways to do such additions.

It is a good idea to identify each strategy by the name of the person who suggested it. This makes the strategies easier to refer to later on.

"Are any methods easier than others for working out in your head?"

The discussion should involve what is meant by an 'easier' method. It is likely that different people will find certain methods easier than others.

Tell pupils that they are now to work in pairs.

"I am going to write down several additions. In each case, work them out in your head. You may only write down the answers. Discuss the methods with your partner."

Write these additions on the board, one at a time.

56 + 24   35 + 36   43 + 51   76 + 48

Ask them to say which of the methods they think is 'easiest' in each case and why. Emphasise that it is possible to use different methods for different questions. This does not matter as long as the chosen method gives the right answer. Possible strategies include:

$6 + 4 = 10$, *so* $56 + 24 = 50 + 20 + 10 = 80$
$56 + 24 = 56 + 4 + 20 = 60 + 20 = 80$
$35 + 36 = (2 \times 35) + 1 = 70 + 1 = 71$
$43 + 51 = 40 + 50 + 3 + 1 = 94$
$76 + 48 = 74 + 50 = 124$

Pupils may well come up with other methods which are equally valid.

Either on this occasion or another carry out a similar activity with subtraction. We suggest these examples:

80 – 37
58 – 23
101 – 3
84 – 49
72 – 46
31 – 27
100 – 46
300 – 92

Pupils are likely to find subtraction more difficult than addition and are likely to suggest a wider variety of methods to tackle them.
For example:

80 – 37 =
80 – 30 – 7 *or*
80 – 40 + 3 *or*
$37 \xrightarrow{+3} 40 \xrightarrow{+40} 80$
= 43

300 – 92 *causes problems with decomposition, i.e.*

$$\begin{array}{r} \overset{2}{\cancel{3}}\,{}^{1}\overset{9}{\cancel{0}}\,{}^{1}0 \\ -\ \ 9\ 2 \\ \hline 2\ 0\ 8 \end{array},$$

which may not be the best way to do it. Emphasise that written methods of decomposition are often not the best ways for mental calculation, and in many cases mental methods can be more efficient.

The rules of the games 'Make 10, 100, 1000, ...' and 'Reversing' are given on MI 1. You will probably have to explain these rules to pupils, but you may also wish to give them a copy for reference. The games may be played with pencil and paper rather than a calculator, in which case the number chosen does not have to be less than seven. This alternative probably requires less mental effort as you can deal with the question digit by digit, rather than working out and remembering the complete number to enter on the calculator.

MI1

Make 10, 100, 1000, ...

This is a game for two.

You need a calculator. You may only use pencil and paper to keep the score.

Take it in turns to pick a number less than seven. This number gives the number of points you can score and the difficulty of the question you must solve to get them, as described below.

Your opponent enters a number on the calculator with this number of digits. So if you chose the number 2, then your opponent could enter '54', for example.

You take the calculator and, with one addition, you try to make one of the numbers 10 or 100 or 1000 or ... For example, if the original number is 54 then you might enter '+ 46' and press '=' to show '100'.

If you succeed you score your points, in this example 2.

Have five turns each. The person with the higher score wins.

Reversing

This is a game for two.

The rules are the same as for 'Make 10, 100, 1000, ...' but this time the aim is to reverse the digits of the number entered by your opponent. So if you select 3 and your opponent enters '254', you have to make the display say '452', by entering '+ 198' and pressing '='. This time you may need to subtract a number instead of adding.

4 Imaginary cards

Introduction This is a purely mental activity, based on imaginary cards, leading to some written work on products.

Possible content Properties of numbers; mental calculations; multiplication tables; last digits of products.

Preamble This activity depends on pupils' willingness to use their imagination.

Explain that you have nine 'cards' in your pocket. The cards are the same size as ordinary playing cards, the only difference being that your cards just have numbers on them and nothing else. On one card there is a 1, on another a 2, on another a 3, and so on, up to and including 9.

Take the cards from your pocket and shuffle them. Fan them out and ask a pupil to take one, look at it, tell the class what number is on it and then keep it.

Repeat the above process with three other pupils. Remember though that all four cards must have a different number.

At this point you can do some work with these four numbers by asking a variety of questions, in particular to those pupils who do not have a card. Such questions might be:

"What is the largest number on the cards chosen?"

"What is the smallest?"

"What is the sum of the numbers on the four chosen cards?"

"Tell me one number that is on the cards left in my hand ... tell me another ... and another."

"Tell me one odd number that is on the cards left in my hand ... and another."

"Tell me one even number ... and another."

Collect in the cards and shuffle again!

To involve pupils in the spirit of this activity, invite one of them to shuffle the pack!

Before moving on to the next stage, you may wish to repeat the above process with a newly selected set of four cards. You could use similar, easier or harder questions depending on the pupils' response.

Ask two pupils to select a card each and to announce to the rest of the class what numbers they have chosen.

"I want all of you, in your mind, to multiply these numbers together. Do not call out your answer."

Ask several pupils to tell you what they think the last digit of the product is.

If the product is a single digit you will have to adjust the question. It is important that you ask several pupils. Agree on the correct answer.

Repeat this with other pairs of cards until the process is understood by the pupils.

Collect in the cards. Shuffle them. Ask a pupil to choose any two cards, keep the numbers secret and then tell the others the last digit of the product.

Ask members of the rest of the group to tell you which two cards they think have been selected.

Discuss the answers received. Some may not be correct. There may be more than one possible solution. Check with the person who selected the cards.

Repeat the process with other pupils choosing pairs of cards.

Establish, through discussion, that for many of the last digits there is more than one possible pair of starting numbers.

Collect the cards and put them away!

Set the pupils the task of determining the last digits of single-number products and finding all the possible combinations for each digit.

You may wish to include the same digit twice e.g. 3×3, 4×4.

Ask them to think carefully beforehand about how they will record the numbers they use and the results they obtain.

Accept all methods. You may wish to discuss the merits of each one later.

Choose two 2-digit numbers and ask what they think the last digit of the product will be and why.

Discuss the responses and the reasons offered. Allow pupils to check their predictions on a calculator. Depending upon the level of understanding arising out of the discussions, you may wish to allow them to investigate this situation further. You could extend the investigation to include both 3- and 4-digit numbers.

5 Imaginary dice

Introduction This is the first of a series of activities designed to help pupils do mental calculations like 13×5 or 15×7. 'Imaginary dice' gives pupils practice at storing integers mentally. Subsequent activities develop this idea for doing calculations like those above.

Materials A six-sided dice large enough to be clearly visible from the back of the class – the dice should preferably have dots rather than digits; possibly sheet MI2.

Possible content Simple whole number arithmetic; the vocabulary and mental visualisation associated with cubes.

Preamble These activities are suitable for a whole class.

Hold up the large dice with your hands covering the dots on the top and bottom faces.

Make sure the whole group can see the dice.

"What do the two hidden numbers add up to?"

Turn the dice round (keeping top and bottom faces covered) so that pupils can see all four uncovered numbers. You may need to rotate the dice more than once so that all pupils can respond.

Accept and discuss answers from the group.

Repeat with a different pair of hidden faces.

It is very important that everyone should eventually realise that opposite faces always add to 7.

Hold up the dice so that one face is visible only to you, and the opposite face can be seen by everyone else.

"What number can I see?"

Establish that the number you can see is '7 – (the number they can see)'
The dice should now be put out of sight for the remaining activities.

In the remaining activities, pupils may find it easier to keep track of their mental images if they close their eyes and remain silent.

"Imagine throwing two dice. The left-hand one shows a five on top; the right-hand one shows a three on top."

Give pupils about 15 seconds to fix the image of the dice in their minds.

"What is the total score on your two dice?"

Make sure all pupils realise that the total is 8.

The next two questions are intended to bring the image of the dice back into focus. They do not require discussion.

"You should still be able to see your two dice.
What is the left-hand dice showing on top?
What is the right-hand one showing?"

It is important to check that pupils can accurately recall their mental images of the two dice.

"Turn the right-hand dice right over. What is the total score now?"

Collect and discuss answers, establishing which is correct.

If pupils have difficulty, you could start afresh at the next section.

"What number is your left-hand dice showing on top now?
What number is your right-hand dice showing on top?"

"Turn the left-hand dice right over."

Give sufficient time for pupils to sort out in their own minds what they should be able to see now.

"What is the total of the two numbers on top now?"

"Now turn *both* dice right over. What is the final total on top?"

Collect and discuss the answers, establishing which are correct.

You may wish to round off this particular sequence with a condensed 're-run' through the whole sequence, with the group keeping their eyes shut and picturing what they 'see' at each stage.

"We started with a five and a three on top, then turned the right-hand one over. What did you see?
We then turned the left-hand one over. What did you see then?
Finally we turned both dice over and got back to where we started – a five and a three."

"Clear your mind and throw the dice again.
This time you get a two and a six."

Allow about 10 seconds.

"Turn both dice right over.
What is the total score on the dice now?"

Establish that the answer is 6.

Various strategies are possible; you could discuss them with the group.

"Imagine throwing the two dice again.
This time *you* choose your own two top numbers."

Allow 15–20 seconds to give pupils time to fix their own choice of top numbers.

"What is the total score? Remember it."

"Turn both dice right over and work out the new total."

"Add this to the previous total. What do you get?"

Collect and discuss answers. Establish that the correct answer is 14.

It may be necessary to repeat this activity.

With some classes this may be an appropriate stopping point.

The next two activities look a little different, but use the same ideas and depend on the same ability to hold integers in separate mental stores.

"Imagine a dice with a five showing on top."

Allow about 10 seconds.

"What do the four numbers round the sides of the dice add up to?"

Allow sufficient time for pupils to:

(a) work out what number is on the bottom,
(b) identify and add up the other four numbers.

They might use different strategies. Collect and discuss answers and methods. Establish that the correct answer is 14.
Repeat, this time allowing pupils to choose their own top number.

"Imagine two dice on a table, both showing a one on top.
Now put one dice on top of the other."

Allow about 15 seconds.

"What is the total of the numbers on the three hidden faces?"

Establish that the correct answer is 13.
(An alternative would be to ask for the total of all the numbers which would be visible if you were allowed to walk round the 'tower'.)

One way of rounding off the activity is to divide the pupils into pairs, with each pair trying one or more of the following, all of which should be done *mentally, without the aid of pencil and paper*, but with plenty of discussion between partners about what they are 'seeing'.

1 The pupils in each pair could take turns to ask each other visualisation problems similar to the ones above. Possible variations might be to use three dice or to use eight-sided dice, if pupils are familiar with them.

2 The pair could tackle the problems on sheet MI2, identifying and agreeing on what the invisible numbers indicated by question marks stand for. Two pairs could then join to form a group of four to compare answers and to discuss the methods they used.

3 The pairs could investigate mentally what are the largest and smallest possible totals of the visible faces for 'towers' of different heights greater than or equal to 2.

Imaginary dice

What numbers are the arrows pointing to?

1

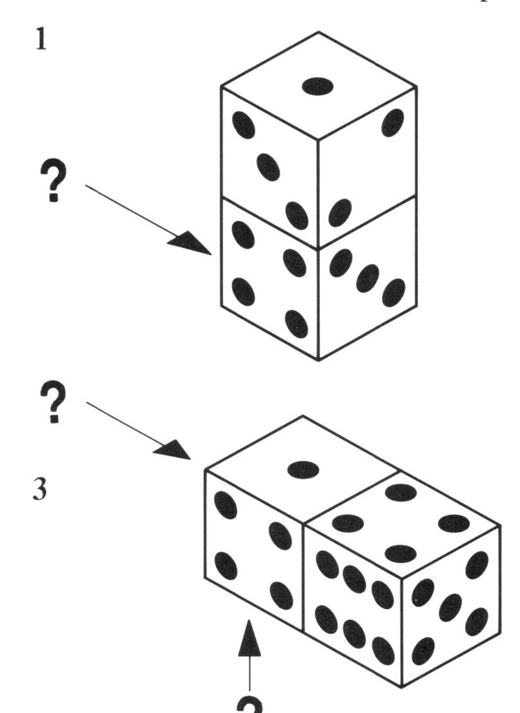

2

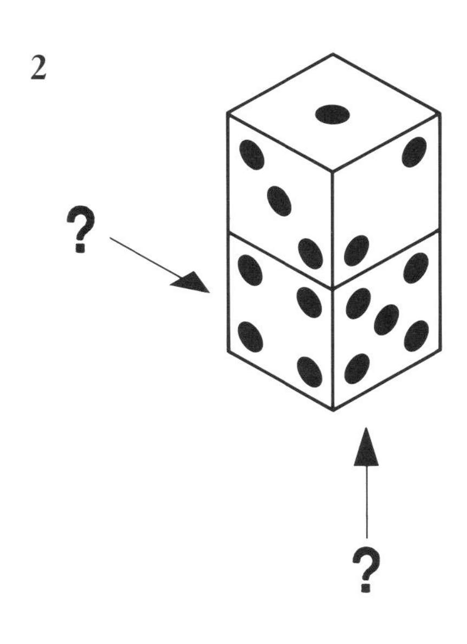

3

4 For this tower of three dice:
(a) What is the sum of all the numbers you cannot see?
(b) What do the six numbers around the back add up to?
(c) What are the five completely hidden numbers?

5 Imagine a normal six-sided dice.
Looking at one of the corners, you can see the faces which are numbered – in clockwise order – 1, 2, 3.

If you went round the other side and looked at the opposite corner, what numbers would you see *in clockwise order?*

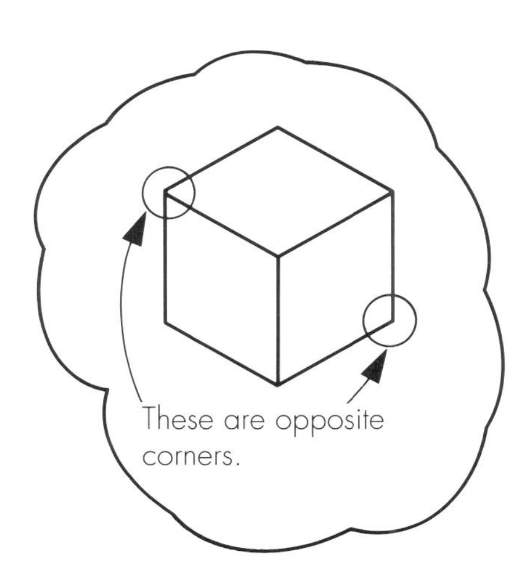

6 Boxes

Introduction The aim of 'Boxes' is to develop pupils' mental ability to store and manipulate numbers. It follows on from 'Imaginary dice'.

Possible content Whole number arithmetic.

Preamble These activities are suitable for a whole class.

"These activities are designed to help you handle numbers mentally.
It can sometimes be useful for you to store numbers in your memory."

"Close your eyes.
Imagine a square box which can store numbers.
At the moment the box is empty."

Allow sufficient time for the image to stabilise.

"I am going to read out a list of numbers.
Keep a total of these in the box.
Each time the total is a multiple of 5 put your hand up for a couple of seconds.

2 6 **7** 4 **1** 6 3 **6** 2"

Read out the list of numbers. The bold numbers indicate where the total is a multiple of 5.

From the 'hands-up' it is possible to judge how effective the group is at this activity. It may be useful to watch out for those pupils who never put their hands up.

"Now open your eyes."

Give some general feedback about how successful people were.

"Close your eyes again and empty the box.
I am going to read out another list of numbers.
Put your hands up whenever the total is a multiple of 5.

5 3 **2** 9 2 3 **1** 6 2 **2** 1 5 8 **6** 3 4
Now open your eyes."

Discuss any problems pupils may have found and strategies they could have used. For example, every time a multiple of 5 is reached you can start counting again from zero. Counting in modulo ten could be used here.
Repeat the activity with fives or other multiples.
It is strongly recommended that a pre-prepared list be used.

"Close your eyes. Imagine two boxes, a left-hand one and a right-hand one."

Allow about 5 seconds.

"These two boxes can both store numbers; at the moment both are empty."

Allow pupils a few seconds to stabilise this image in their minds.

"I am going to read out some numbers. I will tell you which box to add them to."

Allow several seconds between instructions.

"Put 3 in the left-hand box and 5 in the right-hand box.
Now add 4 to the left-hand box.
Add 1 to the right-hand box.
Add 7 to the left-hand box.
Add 4 to the right-hand box.
What is the total in the left-hand box?
What is the total in the right-hand box?"

Collect and discuss the answers.
Establish that there is 14 in the left-hand box and 10 in the right-hand box.

Now repeat the activity using your own numbers.
Prepare the numbers you are going to use in advance. Keep the total for each box less than 20.
Collect and discuss the answers.

"Clear the numbers in your two boxes.
Make sure they are both empty."

"I am now going to read out a list of numbers. Do not add them this time.
I want you to keep track of the smallest and largest of the numbers.
Use one of your mental boxes for the smallest number, the other for the largest."

Allow pupils a few seconds to stabilise this image in their minds.
Allow several seconds between numbers.

"23 8 24 18 20 17 15 29 25 18 9 6 5 29 28."

Collect in the answers quickly and discuss possible strategies which pupils employed.
Repeat the activity with your own numbers.

With some pupils it might be useful to touch upon some ways such a process might be done by a computer.

7 Bigger boxes

Introduction The aim of 'Bigger boxes' is to develop pupils' mental ability to store and manipulate numbers. 'Bigger boxes' follows on directly from 'Boxes'.

Possible content Whole number arithmetic (including the addition of two-digit numbers).

Preamble These activities are suitable for a whole class.

These initial activities involve easy combinations of numbers. The idea is to encourage the use of mental boxes.
If pupils experience any difficulties with the first three of these activities, it indicates that further number work of this kind is required. It would be pointless to continue unless pupils have this basic competency.

Read the numbers slowly.

"These activities are designed to help you handle numbers mentally.
I am going to read out about half a dozen numbers. I want you to add them up. It may help you to close your eyes.
15 5 20 25 10 15"

Quickly establish the correct answer (90).
All these numbers are multiples of 5.

"Now let's try the same thing with these numbers:
18 2 12 10 8 8"

Quickly establish the correct answer (58).
Give some positive feedback.
These numbers all involve $2 + 8$ *type combinations.*

"Now try these numbers:
14 6 10 16 4 10"

Again, quickly establish the answer (60).
These numbers all involve $4 + 6$ *type combinations.*

"Now try these:
15 6 12 5 18"

Give the answer (56).

"Which set of numbers did you find the hardest to add up?"

Allow a fairly open discussion in order to establish the ease of handling totals in ten (that is $4 + 6$, $8 + 2$, $5 + 5$, $7 + 3$*).*
In order to reinforce this, get pupils to set each other additions which involve these 'easy' totals.

The following activities involve 'harder' numbers than in the previous activities.

"Close your eyes. Imagine two two teams playing a game: the red team and the blue team. The first team to reach 50 points or more wins."

"I am going to read out what each team scores.
Put your hand up when a team has won.

Red team scores 12
Blue team scores 6
Red team scores 6
Blue team scores 14
Red team scores 20
Blue team scores 5
Red team scores 15
Blue team scores 25
Red team scores 15"

This should allow you to determine if there are any pupils having difficulty with this sort of activity. Remediation may be further practice or a group discussion concerned with what pupils 'see' or 'feel' during this activity, as well as possible strategies they may employ. It is expected that most pupils will use a 'red' and a 'blue' box, but other more interesting methods may emerge during the course of the discussion.

This activity involves finding the mean of about five or six numbers. Pupils have to count the number of numbers as well as keeping track of the cumulative total. The answers should be integers.

"I am going to read you several numbers.
You will need to keep a check of how many.
When I have finished I want you to work out the mean of these numbers: 4 6 15 2 3"

Establish the correct answer (6).
Encourage pupils to tell of any difficulties they experienced with the task. If necessary repeat the activity several times. Use your own pre-prepared numbers.

"In cross-country competitions the finishing position of a runner gives his or her score.
The team with the smallest score wins the match."

The situation may need further explanation.

"In a cross-country competition there are two teams: the red team and the blue team. I am going to give you the order in which the members of the teams finished. I want you to work out – in your heads – how many points each team scored."

For some groups it may be useful to discuss a possible strategy before embarking on the activity. This must be a matter of personal judgement.

"Here is how the runners finished:

1st	Red
2nd	Blue
3rd	Blue
4th	Red
5th	Blue
6th	Red"

Collect and discuss pupils' answers for the team scores. It may be useful to discuss with the group what, if anything, they found hard about the activity and what they 'saw' or 'felt'. This may be useful in showing the group that there are both similarities and differences between different people's mental pictures.

This may be a useful point to stop. However, the following additional activity might be found useful by some groups.

Tell pupils to imagine two boxes.
Read out a string of numbers, pupils choosing which box to put them in.
The idea is to keep the totals in each box as close as possible.

8 Mental multiplication

Introduction This is the last in a series of activities designed to help pupils perform mental calculations like 13 × 5 or 15 × 7. It is important that pupils have already gained experience in mentally storing numbers in one or two mental boxes. This should have been achieved after the activities in 'Imaginary dice', 'Boxes' and 'Bigger boxes'.

Possible content Mental calculations involving the multiplication of a two-digit by a one-digit number.

Preamble These activities are suitable for a whole class.

It is crucial that the pace is not too fast or only a superficial understanding may result.

Write down the calculation '15 × 7' on the board.

"How many different ways could you use to find the answer to this?"

Collect and record answers.
Allow sufficient time for a variety of 'methods' to emerge.
Some responses might be:

15 + 15 + 15 + 15 + 15 + 15 + 15

30 + 30 + 30 + 15

7 × 5 + 7 × 10

10 × 7 + 5 × 7

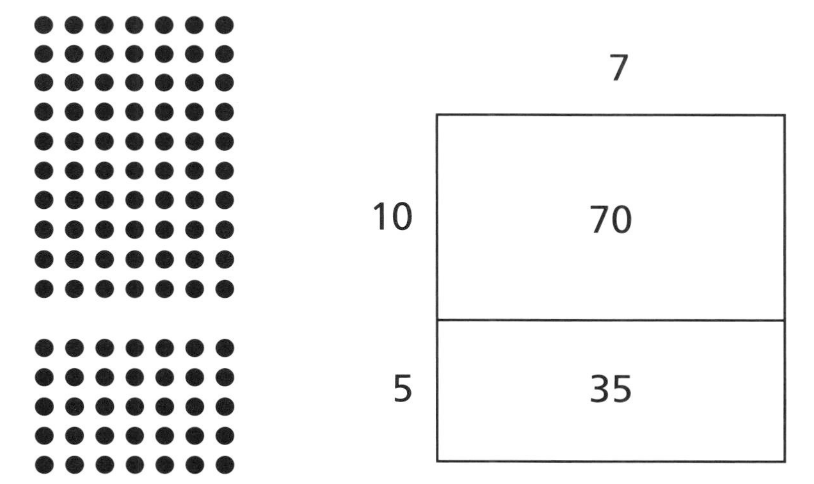

Ensure that '10 × 7 + 5 × 7' *or some variation appears.*

"Which of these ways do you think would be easiest to do in your head?"

Allow a succinct but free-ranging discussion. Do not be judgemental.

"It sometimes helps to picture number boxes. Which of your ways could make use of number boxes?"

Again, do not force the issue. Allow pupils to draw on the board to amplify their mental processes.

If the following method does not emerge, then describe it on the board. Stress that it is a method which many people find useful (largely because it is a general strategy). Their method might well be superior.

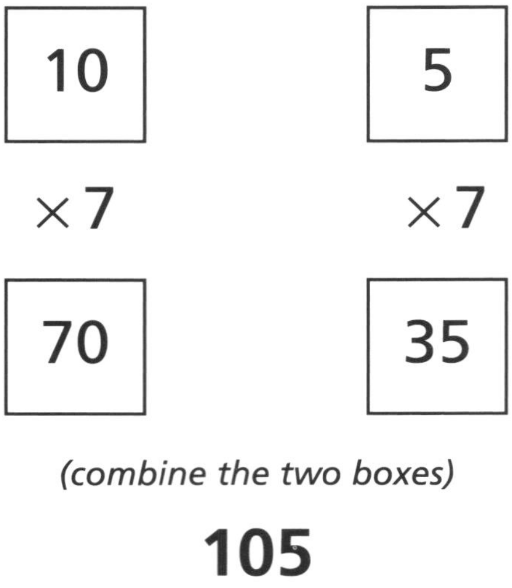

10	5
×7	×7
70	35

(combine the two boxes)

105

It may be found useful if, each time the contents of a box change, the old box is erased.

Look at the other methods from the point of view of generality; for example, 15 × 4 = 30 + 30 *would not generally be applicable.*

"Work out 23 × 5. Use mental number boxes."

Collect answers. Encourage pupils to discuss their methods. Try a few more calculations (such as 21 × 4, 4 × 16, 6 × 15*).*

Get pupils to try mental multiplications on each other.
After this discuss any methods or difficulties encountered by pupils. If possible 'nudge' the discussion into talking about multiplications they find easy. These might include 25 × 4 = 100, 50 × 2 = 100, *etc.*

9 Matador

Introduction This is a game for two to six players. The aim is to encourage visualisation of number bonds to 10. There is the opportunity for players to develop ideas of strategy.

Materials Double-nine set of dominoes. (These are available from good games shops and from E. J. Arnold.)

Possible content Addition of single-digit numbers.

Preamble It is important to ensure that the rules are understood, and to do this it may be best to start with an unscored game where the rules are discussed.

This game should not be played close to 'Fives and threes' because the rules are very different and could be confused.

This game is not like normal dominoes. In 'Matador', to add a domino to the end of a line, the dots must add to 10, like this:

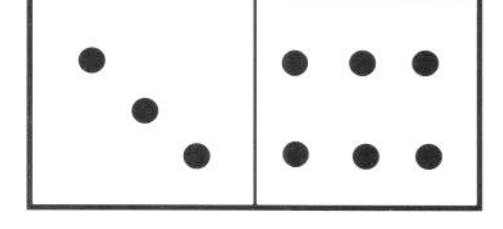
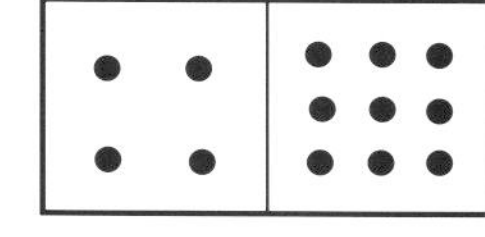
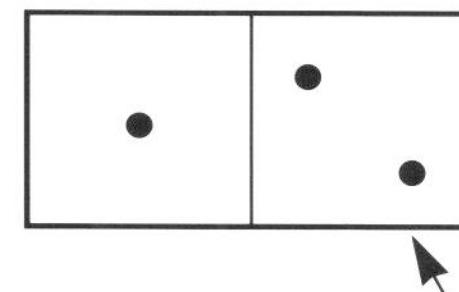
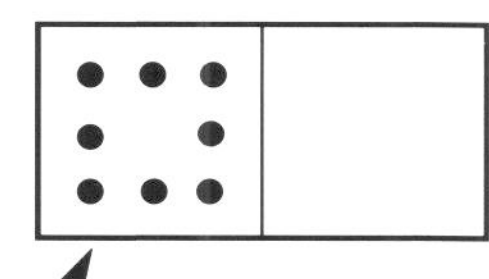

These dots must add to 10.

The dominoes whose dots add to 10 (1:9, 2:8, 3:7, 4:6 and 5:5) and the double blank (0:0) are called matadors. They may be played at any time and they are the only dominoes which may be played next to a blank.
Dominoes are placed in a long line, turning corners as necessary, like this:

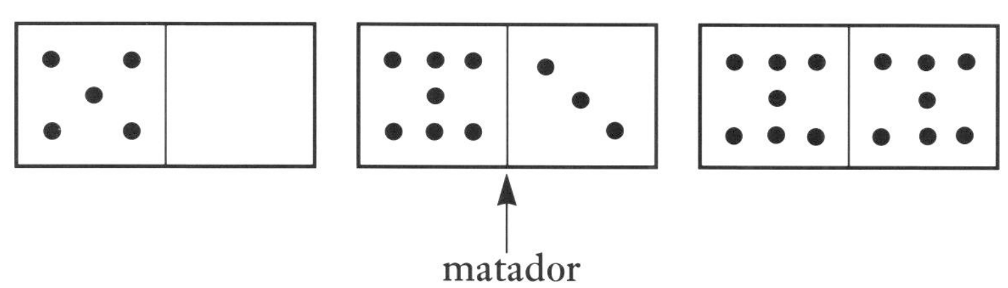

1 **Deciding who starts** The dominoes are all placed face down and shuffled. Each player takes a domino in turn and places it face up. The person with the highest double (or the highest total if no one has a double) starts. The dominoes are all placed face down again and shuffled.

2 **The draw** Each player takes seven dominoes. They may look at their own dominoes but not at anyone else's.

3 **The play** The first player plays a double if possible; otherwise they play their domino with the highest total.
The player on their left plays a domino so that 10 is formed, if they can. If they cannot do this, they may either play a matador or take dominoes from the pool, one at a time, until they can play. (If they have a matador, they can play it at any time.)

Play continues in a clockwise direction. When no dominoes are left in the pool, players must play one of their dominoes if they can. If a player cannot play, they say 'pass' and play passes to the next person.

4 **How the play ends** If a player plays their last domino, they score the total of all the spots on the dominoes held by the others. If no one can play, so that the game is blocked, the player with the lowest total of spots scores the total of the dominoes held by the rest.

5 **The winner** The winner is the first person to score 200.

10 Fives and threes

Introduction This is a set of three games whose aim is to encourage awareness of multiples of 5 and of 3.

Materials Double-nine or double-six set of dominoes. (Double-nine dominoes are available from good games shops and from E. J. Arnold.)

Possible content Familiarity with multiples of 3 and 5; mental addition of small numbers.

Preamble There are three possible games, each for two to four players: 'All fives', 'All threes', and 'Fives and threes'. For each game there is an introductory game which is designed to familiarise pupils with the method of scoring. It is a good idea for pupils to start with 'All fives' and to progress to the other, more difficult, games as they become more proficient.

With a normal, double-six set of dominoes, the numbers are fairly small, but with the double-nine set the game becomes quite interesting and requires a lot of mental calculation for 'Fives and threes'.

These games should not be played too close to 'Matador' because the rules are very different and this could cause confusion.

■ All fives

Rules of introductory game

Put all the dominoes face up. Take it in turns to take a domino. Add together the spots on the domino and score a point for every multiple of 5, so that a total of 5 scores one point, 10 scores two points, etc. Totals which are not multiples of 5 do not score any points.

Continue taking dominoes until all the multiples of 5 have gone. The player with the highest number of points wins.

Rules of 'All fives'

1 **General** As in normal dominoes, dominoes are played in a long line, turning corners as necessary. Doubles are placed at right angles. A domino may only be laid down if one of its ends matches a free end. Points are scored during play every time the total of spots on the free ends is a multiple of 5.
The easiest way to keep score is to use counters on a number line.

2 **Deciding who starts** Take it in turns to start.

3 **The draw** The dominoes are all placed face down and shuffled. Each player takes five dominoes.

4 **The play** The first player can play any domino. Add the spots. A total of 5 scores one point, 10 scores two, and 15 scores three.
The player on the left plays next. They may play any domino which matches one of the free ends and they must play if they can. If they are unable to play, then they keep taking dominoes, one at a time, from the

pool until they can play. When they play a domino, they score if the spots at the ends of the line of dominoes add up to a multiple of 5. If an end domino is a double, all the spots on that domino are counted as an end. The score is one if the total is 5, two if the total is 10, three if the total is 15, and so on.
Play continues in a clockwise direction. If a player cannot go and there are no dominoes left in the pool, they must just say 'pass'.
When a player has laid down a domino, they must state their score even if it is zero. If a player fails to claim their score, the first player to notice can claim the points.

5 **How the play ends** Play continues until one player has no dominoes left or all players have passed and the game is blocked. If a player plays all their dominoes they have a bonus of three points.

6 **The winner** The winner is the first person to score 50 points.

All threes

This is played in the same way as 'All fives', but points are scored for multiples of 3 rather than multiples of 5. For example, a total of 6 scores two points.

Fives and threes

This is played in the same way as the previous games, except that points are scored for both multiples of 3 and multiples of 5. For example, 10 scores two (2 × 5) and 12 scores four (4 × 3), but 15 scores eight (**3** × 5 and **5** × 3).

Challenge

What is the highest score you can get in a single move? In what ways can you get it?

11 Dominoes

Introduction This is a selection of activities which use mental arithmetic.

Materials Double-six and double-nine dominoes (available from good games shops and from E. J. Arnold); sheet MI 3.

Possible content Mental arithmetic.

Preamble These activities are all fairly short and need not be attempted in the order given. It is intended that one or two should be selected for use at any one time.

Activities 2–4 should only be attempted when pupils are familiar with dominoes. If they are not, then activity 1 should be done first, using dominoes.

Hand out the pupil activity sheet MI3. Pupils should be able to manage it on their own, but take the following points into consideration.

Activity 1
This activity should be done with dominoes if pupils are not familiar with them. If it is done without, then it is more of a mental activity.

Activity 3
If pupils have problems, they could be allowed to do one ring using actual dominoes.

Dominoes

1 Choose combinations of dominoes where the total number of spots is 20.
Is this possible with two dominoes, three dominoes, four dominoes, etc.?
What is the largest number of dominoes you can find with spots adding to 20 altogether?

2 Place any domino face up on the table.

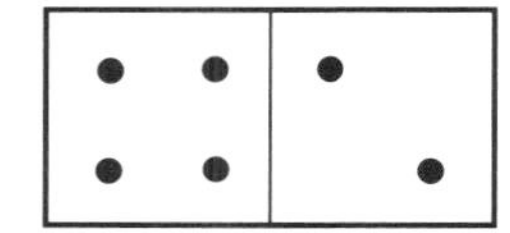

Add together the two end numbers. (In this case, the total is 6.)

Now place a second domino end-to-end with the first, so that the two ends next to each other match and the total on the free ends is one more than it used to be. In this example, there are two possibilities:

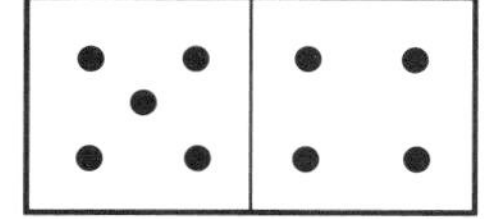 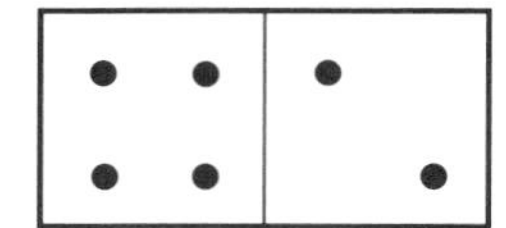 and 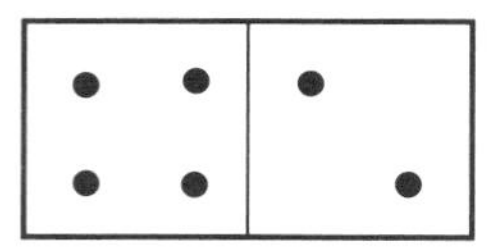 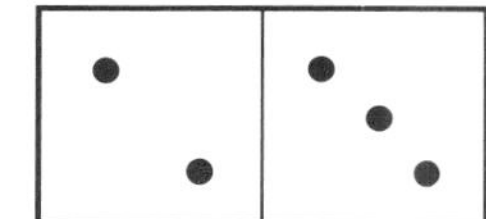

Carry on the sequence by placing a third domino on either end of your line so that the total of the free ends is one more (8 in this case).

What is the longest line you can have?
Does it matter which domino you start with?

3 Imagine that four dominoes are placed in a ring like this, so that ends touching each other have the same number of spots.

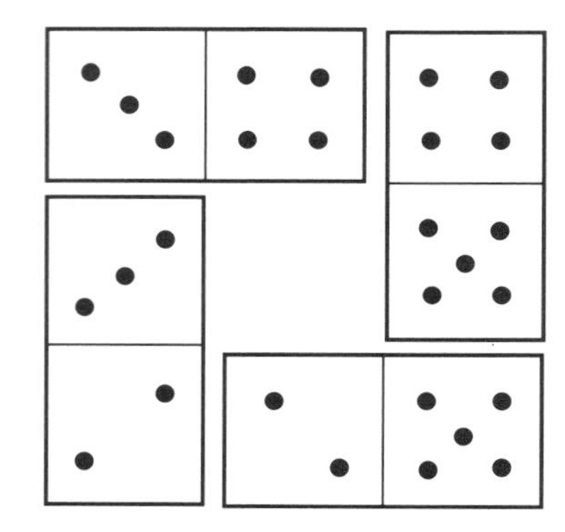

What is the smallest possible total of spots? What is the largest possible total? Is it possible to make every total between the smallest and largest number? If not, why not?

Test your answers using dominoes.

4 Think of three dominoes so that the total numbers of spots across and down are as shown. Ends touching do not necessarily have to be the same.

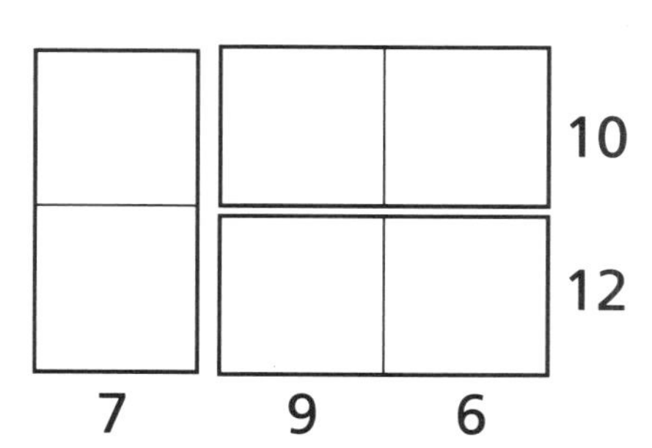

Are there any other groups of three dominoes which would also work?

Check your answers using dominoes.

12 Posts and gaps

Introduction This activity helps pupils use visual imagery to analyse a common type of word problem in which the answer is often one more or one less than expected.

Possible content Mental arithmetic.

Preamble When objects are strung out in a line, the number of gaps *between* the objects is one *less* than the number of objects.

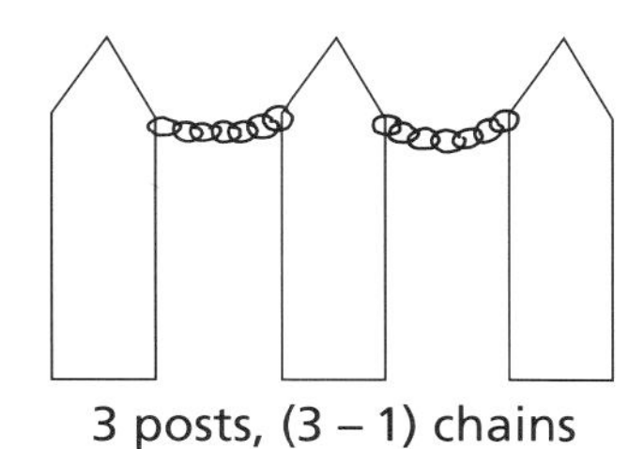

3 posts, (3 – 1) chains

If the two end gaps are counted, the number of gaps is one *more* than the number of objects.

3 window bars, (3 + 1) gaps

Care is needed in deciding which situation is applicable to a given problem.

The first problem raises the main issues which this activity addresses and needs careful thought.

"I am going to pose a problem which I want you to do in your heads. I will state the problem *twice only*."

"Six bus stops are equally spaced along a bus route. The distance from the first to the third is 600 m. How far is it from the first stop to the last? Explain how you got your answer."

Collect answers and ***explanations*** *from pupils without discussing them at this stage.*
You are likely to get a range of responses (some of which may produce an incorrect solution) such as:

1st to 6th = 2 × (1st to 3rd) = 1200

600 = 3 × ?, so ? = 200; therefore 1st to 6th = 6 stages = 1200

1st to 3rd = 2 stages, so 1st to 6th = 6 stages = 1800

1st to 3rd = 600, so 1 stage = 200; therefore 5 stages = 1000

600 = 2 × ?, so 5 × ? = 1500

1st to 3rd (miss 1) = 600, so 1st to 6th (miss 4) = 2400

Now get pupils to draw a diagram representing the bus route. Use the diagram when discussing the explanations.

Stress the key feature of this problem which the diagram brings out:

6 bus stops → (6 – 1) gaps

1st to 3rd → (3 – 1) gaps

Establish the correct solution (1500m).

"I am going to give you some problems to do in your head. As I read out each problem, I want you to make up a mental picture, like the picture you drew of the bus route, to help you. I shall then pause before asking the question. Think carefully before you write down the answer."

1 "I want to erect a fence 10m long at the end of my garden. The panels are 2m wide and every panel has to be supported by a post at each end." (*Pause.*)

"How many posts do I need?"

2 "Chapter 7 of *Why Johnny can't add* starts on page 74 and finishes on page 83." (*Pause.*)

"How many pages are there in the chapter?"

3 "My prison cell window has three vertical bars." (*Pause.*)

"How many gaps does it have?"

4 "Houses down one side of the road are numbered 1, 2, 3, up to 40." (*Pause.*)

"How many houses are there in between number 17 and number 35?"

5 "I want to cut a metre rule into 10cm sticks." (*Pause.*)

"In how many different places must I cut it?"

6 "A school sells numbered dinner tickets. Yesterday the first ticket sold was number 265. Today the first ticket sold was 414." (*Pause.*)

"How many tickets did the school sell yesterday?"

7 "I have to hang out ten pillowcases to dry. If I use two clothes-pegs for each pillowcase, I will need twenty pegs altogether. Instead of doing this, I use the same peg for the right-hand corner of one pillowcase and the left-hand corner of the next." (*Pause.*)

"How many pegs will I need?"

8 "I count from 30 up to 100." (*Pause.*)

"How many numbers do I count?"

Now go back over each question in turn and discuss answers with the class, and the mental images they used to obtain their answers. Bring out the structure in each solution rather than the answer itself. Stress the fact that these problems are of two distinct types:

(a) problems 1, 3, 5 and 7 are similar in that they involve unnumbered physical objects which need to be imagined in order to see what is required, and

(b) problems 2, 4, 6 and 8 are similar in that they involve numbered objects, though again you have to think carefully to see exactly what to do with these numbers in order to get the required answer.
For example:

1 $10 \div 2 = 5 \to 5$ panels, so $5 + 1$ posts

3 3 bars $\to$ ($3 - 1$ gaps between) + 2 end gaps = $3 + 1$ gaps

5 1 m = 100 cm $\to$ 10 bits $\to$ $10 - 1$ cuts

7 10 pillowcases $\to$ ($10 - 1$ joins) + 2 ends = $10 + 1$ pegs

Similarly:

2 Chapter 7 contains all the pages up to 83 *except for* the first 73 pages $\to 83 - 73$.

4 Houses between 17 and 35 $\to$ all houses up to 34 *except for* numbers 1 to 17 $\to 34 - 17$.

6 Tickets from 265 and before 414 $\to$ all up to 413 *except for* the first 264 $\to 413 - 264$.

8 Counting from 30 up to 100 counts all numbers up to 100 *except for* the first 29 $\to 100 - 29$.

Now that pupils are aware of the pitfalls, get them to tackle the following problems.

9 "I have 25 fence panels, each 2 m wide, and 24 fence posts. Every panel has to be supported by a post at each end. How long a fence can I erect?"

10 "Bus stops are equally spaced along a 6 km bus route. The distance from the first to the fourth is 1·2 km. How many stops are there altogether?"

11 "Yesterday evening in bed I read from page 67 to page 123. How many times did I turn a page over?"

12 "In a 110 metre hurdle race the first hurdle comes after 15 metres and the last hurdle comes 15 metres from the end. There are nine hurdles. How far apart are they?"

13 "I have a 20-volume encyclopaedia on a special shelf of its own. When my daughter borrows a volume for her homework, she replaces it the right way round but puts it anywhere on the shelf. In how many different places can she put it back?"

14 (a) "Suppose I have a chocolate bar with 6 pieces in a single row. How many breaks must I make to get 6 separate pieces?"

(b) "Suppose I have a 6 by 4 block of chocolate. How many breaks must I make to get 24 separate pieces? Is there a best way?"

15 "How many points are there on a chessboard where four squares meet?"

16 "One day I tidy my encyclopaedia so that all 20 volumes are in the correct order. When I next use it, I discover a bookworm has eaten its way from the front cover of volume 1 to the back cover of volume 20. How many volumes has the bookworm chewed through?"

13 Mind measures

Introduction The aim of these activities is to help pupils become aware of the various processes involved in estimating length and area.

Materials A stick 10 cm long; metre rule; sheet MI4 on an OHP transparency.

Possible content Estimating lengths and areas.

Preamble It is important to allow pupils sufficient time to reflect upon their own and others' strategies for length and area estimation. Estimates of length will be more accurate than estimates of area.

You will have to identify two points 60 cm apart horizontally and perhaps some other points whose distances apart are known – it may be helpful to have these prepared before the lesson.

Show sheet MI4 on the OHP.

"On this sheet are some patterns made out of triangles or squares.
I want you to tell me how many of the black shapes are needed to cover each shape."

Write down – with no comment – all the answers.
Then ask for the methods employed and discuss the answers obtained.
The question of whether or not to include the black shapes may arise at this point.

"For the next activity you are going to use a ruler in your mind to estimate some lengths.
This stick is 10 cm long. Fix the length in your mind."

Hold up the 10 cm stick so that all can see its full extent clearly.

Allow sufficient time, perhaps 5 seconds, for pupils to fix the image in their minds.

Indicate two points about 60 cm apart horizontally.

"How many centimetres is it from here to here?"

Write on the board all the answers as they are given – with no comment.

"How did you estimate that distance?
What method did you use?"

Encourage pupils to talk about their strategies, for example stepping off or using other background or personal measures. Ensure that all pupils have a chance to share their own particular method with the group.
Finally ask a pupil to measure the distance with a ruler.

Repeat the estimating activity as appropriate, using different horizontal or vertical distances.

"Why is it helpful to have your own mental ruler when you have to estimate lengths?"

Guide the group towards realising that an internal standard makes stepping off easier to carry out. Establish that, once you have a mental yardstick, estimation of length only involves counting.

"Estimate the length of this room."

Collect on the board the answers from the whole group as they are given – with no comment. Then ask a pupil to check by measuring with a metre rule. The intention of this activity is to make the group realise that the 10cm mental ruler is not always appropriate.

"When would it be more useful to use other mental rulers? What mental rulers do you use? You probably use more than one."

Collect and write these on the board with no comment.
Discuss these mental rulers/yardsticks and bring out the realisation that more than one mental ruler is needed for a wide range of length estimation. For general use, three are needed:
for short (≈ 1cm or 1 inch), medium (≈ 1m) and long (≈ 100m) distances.

Possible extension

- Are people better at estimating lengths which are vertical, horizontal or diagonal?
  This might entail some discussion as to how best to quantify accuracy.

Indicate a wall.
"Estimate the area of this wall."

Collect in the estimates without any comment. Then ask pupils to explain how they arrived at their estimates. Someone is almost certain to have estimated the two dimensions and then multiplied them together – if not, suggest it yourself!

Ask someone to measure the length and height of the wall and use these figures to calculate the wall's area. This may provoke some discussion about units and how accurately the wall should be measured. Discuss this only if it arises naturally.

It is almost certain that the estimates of the wall area will vary considerably. If not, or if it is thought the group could profit from more area estimation activities, ask the group to find the area of a newspaper and/or a sheet of A4 paper. Again, in all probability, the estimates will be far less accurate than the previous estimates of length. Confront the group with this.

"Why do you think your estimates of area are not as accurate as your estimates of length?"

Allow discussion, making a note on the board of the more salient points. When the discussion appears to be virtually at an end consider some concrete examples.

"A wall is nine metres by eight metres. What is its area?"

Record the answer.

"Imagine someone estimated the height and width of the wall. They were accurate to within a metre of the actual lengths. What would be the smallest and largest possible estimates of the area of the wall?"

Some pupils may need help here. However, allow some discussion leading up to the answers 56 square metres and 90 square metres. This should surprise pupils and may need further discussion.

This might be a convenient point at which to stop the activity. However, it could be extended to consider volume. Given a cardboard box, for example a cereal box, it is hard to estimate its volume within an order of magnitude.

You may also wish to estimate mass and time. For angle estimation see 'Watch the angle'.

Mind measures

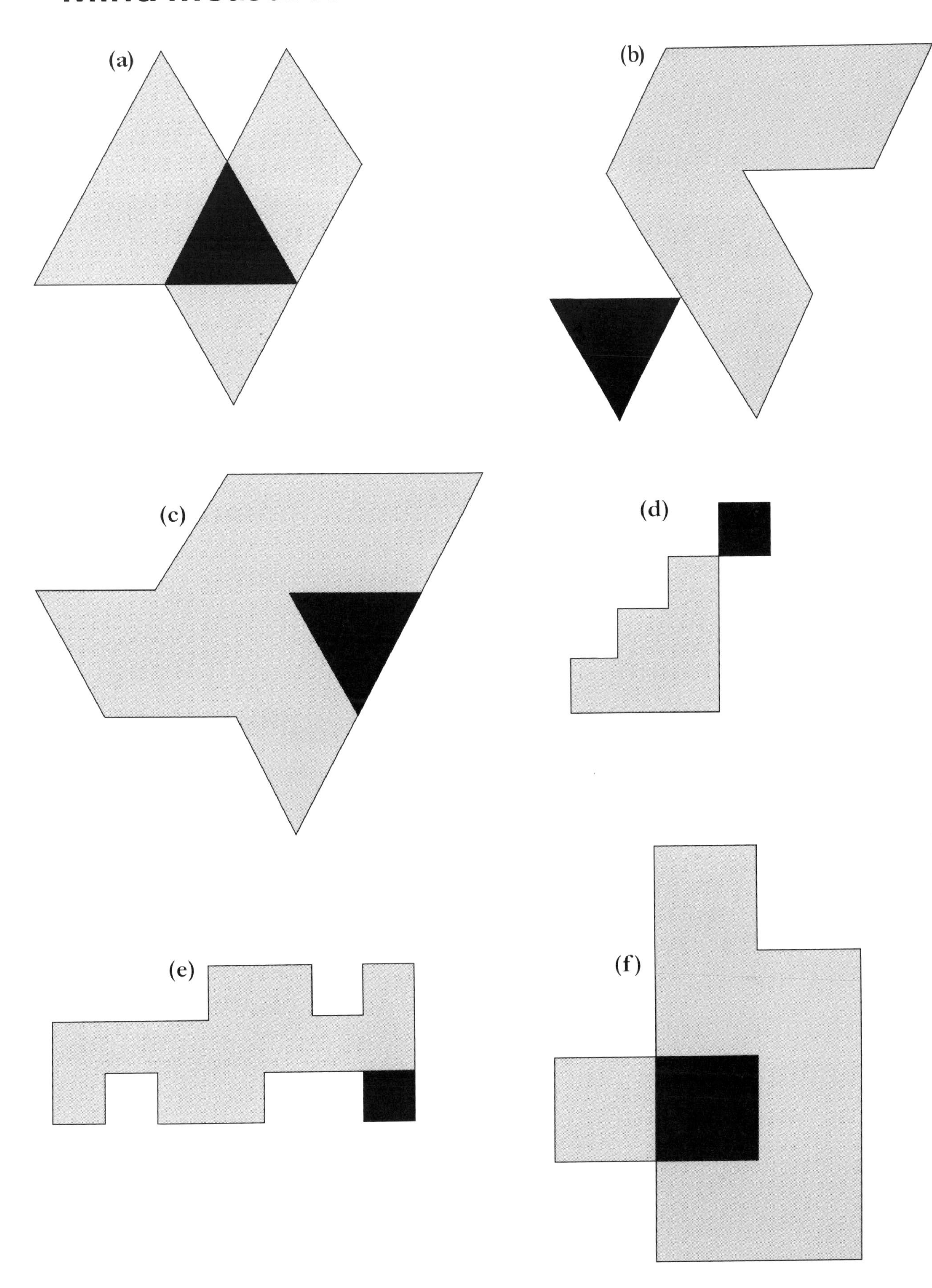

14 Footprints

Introduction The context of a giant humanoid is used to develop pupils' estimation skills.

Materials The *Footprint* poster.

Possible content Ratio; scale factors; estimation.

Preamble This activity is suitable for the whole class. For part of the time, pupils will be working in groups of three or four, which should be organised beforehand.

Ask the pupils if they know what an anthropologist is.

You may have to tell them.

"News has just come in that some anthropologists have found evidence of a humanoid creature. The bones have convinced them that the creature was in human form, but larger than any others known before. Part of the evidence is a fossilised footprint and they have sent this full-size picture of it."

Show the pupils the footprint poster and make sure they realise it is full size.

"Imagine that you are newspaper reporters. Your editor is very excited and wants a front-page story. You have been asked to write the report. In your group, make a list of questions you would want to ask the anthropologist. Decide what information you might want for your newspaper article."

The aim at this stage is to stimulate pupils' imaginations and to encourage them to think of attributes other than height and weight.

At an appropriate time, get the groups to share their ideas. Encourage each group to make notes of questions not on their own list.

It may be necessary to add some of the following ideas to widen the variety.

"Could it step over you / a bus / the school?"

"Could it pick you up in one hand?"

"How long would its stride be?"

"If it lay down, would it fit a swimming pool / hockey pitch?"

"How many eyes did it have?"

"Would its eye be as big as a golf-ball / football / beach-ball?"

"How many of you would weigh the same as it?"

"How much blood would it have in its body?"

"How thick would one of its hairs be?"

"How many ribs would it have?"

"If you drank one can of cola, how much would it need to drink?"

"What animal is as small compared with you as you are compared with the creature?"

Explain to the pupils that whilst they were compiling their questions news has come through that the anthropologists went back to collect all their belongings and have since disappeared. The only evidence anyone has is the picture of the footprint itself.

"In your groups, decide which questions from your list can be answered."

"Think very carefully how an anthropologist might set about determining exactly the attributes of this creature when the only clue they have is the size of the footprint. What methods might they use?"

Ask the groups to compile 'answers' to the questions raised and to establish a clear and agreed picture of what, in their opinion, this creature looked like.

Let each group share its ideas with the others.

At this stage you could ask pupils to write up their report and even draw a picture of what they think the creature looks like. The results could be displayed pictorially or written up as an article with headlines and illustrations. Newspaper reports are often used in other curriculum areas, so it may be worth consulting with relevant colleagues in your school.

Allow the pupils to comment on the advantages or disadvantages of the methods suggested. Hopefully, the suggestion that a comparison of the ratio of their heights, weights, sizes, and so on, to their own feet and the relationship to the creature's foot will be forthcoming.

15 Patterns with numbers

Introduction The aim of this activity is to show that seeing a pattern makes counting easier.

Materials An OHP; multilink cubes or centicubes or counters.

Possible content Factors; multiples; mental multiplication of small whole numbers.

Preamble The activity starts by exploring simple patterns where the number of objects is instantly recognisable. It moves on to more complex patterns where the patterns can be remembered and a calculation needs to be done to find the number of objects shown.

Spread 6 multilink cubes (or centicubes or counters) randomly on an OHP and cover while you switch the OHP on. Uncover for one second.

"How many cubes are there?"

The time allowed for viewing should be short so no one has time to count the cubes.

Arrange 5, 6 and 7 cubes in various ways, some random and others in patterns such as:

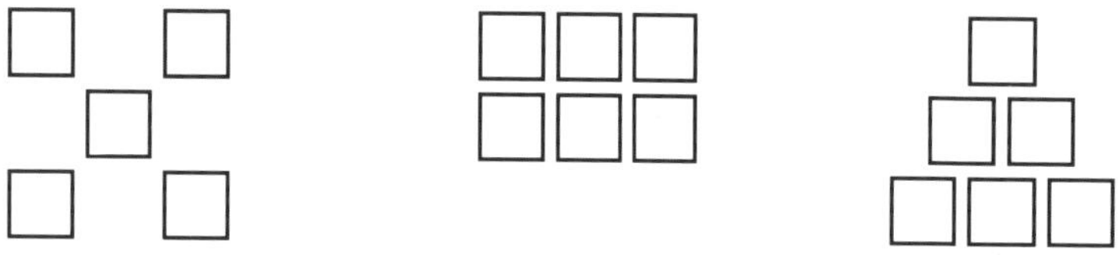

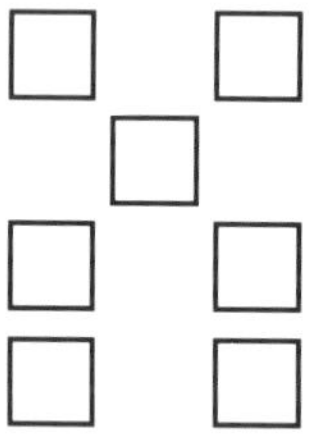

With each arrangement, ask how many cubes there are.

The purpose is to convey the idea that a pattern is helpful.

Spread 12 cubes randomly on the OHP. Uncover for one second.

"How many cubes are there?"

Rearrange the cubes in another random arrangement and ask again.

Then try a more regular arrangement such as:

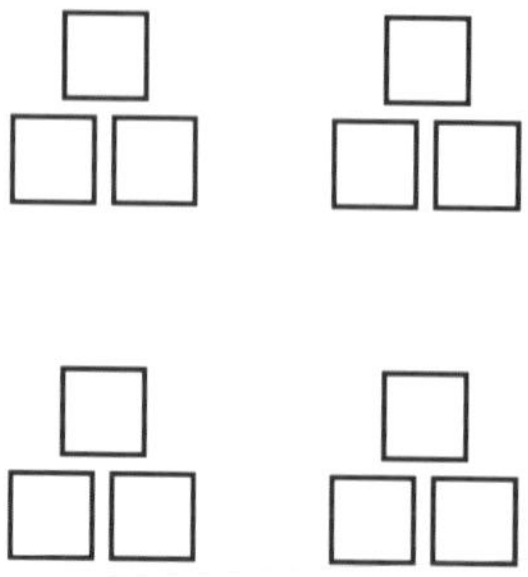

You want to show that a complex pattern can be remembered even though the exact number is not known, and the number of cubes found by calculation from the remembered pattern.

Show other ways of laying out 12 cubes such as:

Discuss which arrangements make recognition easiest.

Now ask pupils to work in pairs.

"Try to find ways to lay out 15, 24, 29, . . . so that people can recognise them easily, even if they only catch a short glimpse."

Encourage pupils to try out their arrangements on each other. This activity should bring in number work on factors and combinations of numbers.

"What is the largest number that you can lay out so that other people can recognise it?"

Recognising a 10 × 10 arrangement in one second may be difficult because of the time needed to count, but recognising four blocks of 5 × 5 may be easier.

You may wish to discuss whether certain patterns are easy to recognise immediately, such as the patterns of dots which appear on dice or playing cards. These patterns can be combined to aid recognition.

You could ask pairs to try their largest numbers out on the rest of the class using the OHP.

Possible extension

- What is the highest number that most people can recognise if the cubes are arranged randomly?

16 Remembering numbers

Introduction This activity encourages pupils to hold small sets of numbers in their memories before answering simple questions about them. You may wish to do 'Square dance' first.

Possible content Factors; multiples; squares; square roots; prime numbers; means; mental arithmetic.

Preamble These suggestions should be used as a series of one-off activities rather than as one long exercise. The opening explanation will probably be needed the first few times this activity is used. Each question could be repeated several times with similar sets of numbers, with larger numbers or with more numbers. The activities could be used with a whole class, group or individuals. It is advisable to write down the numbers first. In selecting which questions to ask, it is important that you are sensitive to the knowledge and abilities of the pupils. For some pupils, it may be appropriate to start with only three numbers.

"I am going to read out a few numbers. Then I am going to ask you some questions about them. After I have read the numbers I will pause for you to fix the numbers in your memory. Do not write the numbers down. When I ask the question, answer it from the numbers in your memory then jot down the answer."

"6 19 3 11 What is the largest number?"

Ask several people what number they wrote down. Ask how many agreed or disagreed. If there is disagreement, repeat the numbers. Some people will always agree with the majority, whatever they wrote down. For them it may be more appropriate to ask similar questions individually or in a small group.

"12 22 5 9 Which is the smallest number?"

"14 21 26 5 Which are even?"

"8 15 12 20 Which are multiples of 4?"

"5 2 7 3 6 Which are factors of 12?"

"34 12 24 9 Which is closest to 19?"

"2 7 9 3 What is the sum of these numbers?"

"34 4 12 9 Which are bigger than 10?"

"2 5 3 7 Which two add up to 10?"

"4 7 3 2 What is the mean?"

"9 27 11 16 Which are square numbers?"

"3 11 10 15 Which are triangle numbers?"

"21 12 17 26 Which is a prime number?"

"23 36 17 25 Which is the largest number?
What is its square root?"

"3 $^{-}4$ 2 5 Which is the negative number?"

"$^{-}5$ $^{-}3$ 4 $^{-}7$ Which are less than negative 4?"

"2·3 4 3·6 3·4 Which are bigger than 3·5?"

"$\frac{3}{5}$ $\frac{2}{7}$ $\frac{4}{9}$ Which is bigger than a half?"

It is possible to have sequences of questions, such as:

"5 8 2 6 Put these numbers in order.
Multiply together the smallest and largest.
Subtract the other two."

17 Imaginary patterns

Introduction This activity is designed to make pupils aware that recognition of pattern can aid memory.

Materials Sheets MI5 and MI6. In some cases it might be useful to have one or both of these copied onto OHP transparencies.

Possible content Simple pattern recognition and explanation; the elementary vocabulary of position and movement.

Preamble These activities are suitable for a whole class.

The first one or two could be done as a large group and, once the general idea is understood and confidence built up, in pairs.

Show the group sheet MI 5.

"I want you to look carefully at this number grid for about a minute."

Indicate the number grid labelled A.
Allow about a minute for pupils to look at the grid.

Cover the sheet / OHP transparency.

"Close your eyes and try to bring a picture of the number grid into your mind. I am going to ask you some questions. When you think you have the answer put your hand up. Don't open your eyes – I'll ask you to answer by name."

"Imagine the square with the number 2 in it."

Allow a second or so for the image to stabilise.

"Which number is directly below it?"

Ask several pupils by name. It is crucial that pupils are given sufficient time to visualise their answers.

Depending on the general ability of the group, a selection of these (or your own) questions may be asked:

"What number is below and to the left of 2?"

"What number is above the number 73?"

"What number is at the bottom left-hand corner of the grid?"

"Which eight numbers surround 78?"

Once pupils know what is expected, the second number grid (B) may be used in the same manner. However, this time ask pupils how *they worked out their answers. The answer that pattern helps should emerge (perhaps after a little encouragement).*

The activity can be terminated or continued with the whole group or continued in pairs – each member 'testing' the other with a different number square or triangle of the type on MI6.

MI 5 Imaginary patterns 1

A

1	2	3	4	5	6	7	8	9	10
11	12	13	14	15	16	17	18	19	20
21	22	23	24	25	26	27	28	29	30
31	32	33	34	35	36	37	38	39	40
41	42	43	44	45	46	47	48	49	50
51	52	53	54	55	56	57	58	59	60
61	62	63	64	65	66	67	68	69	70
71	72	73	74	75	76	77	78	79	80
81	82	83	84	85	86	87	88	89	90
91	92	93	94	95	96	97	98	99	100

B

0	1	2	3	4	5	6	7	8	9
10	11	12	13	14	15	16	17	18	19
20	21	22	23	24	25	26	27	28	29
30	31	32	33	34	35	36	37	38	39
40	41	42	43	44	45	46	47	48	49
50	51	52	53	54	55	56	57	58	59
60	61	62	63	64	65	66	67	68	69
70	71	72	73	74	75	76	77	78	79
80	81	82	83	84	85	86	87	88	89
90	91	92	93	94	95	96	97	98	99

C

1	2	3	4	5	6	7
8	9	10	11	12	13	14
15	16	17	18	19	20	21
22	23	24	25	26	27	28
29	30	31	32	33	34	35
36	37	38	39	40	41	42
43	44	45	46	47	48	49
50	51	52	53	54	55	56
57	58	59	60	61	62	63
64	65	66	67	68	69	70
71	72	73	74	75	76	77
78	79	80	81	82	83	84

D

1	2	3	4	5	6	7	8	9	10	11
12	13	14	15	16	17	18	19	20	21	22
23	24	25	26	27	28	29	30	31	32	33
34	35	36	37	38	39	40	41	42	43	44
45	46	47	48	49	50	51	52	53	54	55
56	57	58	59	60	61	62	63	64	65	66
67	68	69	70	71	72	73	74	75	76	77
78	79	80	81	82	83	84	85	86	87	88
89	90	91	92	93	94	95	96	97	98	99

Imaginary patterns 2

					1					
				2		3				
			4		5		6			
		7		8		9		10		
	11		12		13		14		15	
16		17		18		19		20		21

1									
2	3								
4	5	6							
7	8	9	10						
11	12	13	14	15					
16	17	18	19	20	21				
22	23	24	25	26	27	28			
29	30	31	32	33	34	35	36		
37	38	39	40	41	42	43	44	45	
46	47	48	49	50	51	52	53	54	55

18 Odds and evens

Introduction This activity is designed to help pupils generalise sequences starting from a familiar base, namely odd and even numbers.

Possible content Sequences.

Preamble This activity is good preparation for 'Next please'.

On the board write:

1st	2nd	3rd	4th	...th
2	4	6	8	...

"What number will be next in each row? Why?"

Establish and agree on the correct answers and reasons.
[Do not add the 5th term = 10 to the sequence on the board. The eventual aim of the activity is to emphasise the relationship between *the rows (doubling) rather than* along *the bottom row (adding 2).]*

"What numbers would you expect to come after that? And after that? Why?"

Establish the correct answers and reasons.

Now write '20th' in the top row with '?' underneath.

1st	2nd	3rd	4th	...	20th	...
2	4	6	8		?	

"What number should go in the bottom row? How do you know?"

Establish and agree on the correct answer and reason.
Write '40' in the bottom row.

Tell the pupils you are thinking of a position in the top row but you are not going to tell them which position it is in. Refer to it as the 'cloud' number. Represent it like this:

1st	2nd	3rd	4th	...	20th	...	☁	...
2	4	6	8		40		?	

Ask them to explain to you how they would work out what the bottom number is.

Accept all contributions: cloud add cloud ... cloud times 2 ... 2 times cloud ... double the cloud, and so on. Discuss the solutions offered.

Working with clouds rather than letters is a useful first step to working with an unknown.

When you have agreed upon the solution, place a cloud in the bottom row with a question mark above it.

1st	2nd	3rd	4th	…	20th	…		…	?	…
2	4	6	8		40		?			

Ask how you could work out what number should go on the top row.

Again, accept and discuss all contributions.

The next example is more challenging.

On the board write:

1st	2nd	3rd	4th	…th
1	3	5	7	…

Repeat the above procedure, continuing with:

1st	2nd	3rd	4th	…	20th	…
1	3	5	7		?	

1st	2nd	3rd	4th	…	20th	…		…
1	3	5	7		39		?	

This often leads to a wider variety of responses for the generalised statements. For example, when you place the 'cloud' in the top row, typical responses are:

cloud add cloud subtract 1

2 times cloud subtract 1

(cloud minus 1) add cloud

cloud add (cloud plus 1) subtract 2

One approach is to record all the suggested generalised statements without comment and then ask the pupils to evaluate each one. It may be possible to discuss how and why different-looking expressions are equivalent … an opportunity to introduce some early algebraic manipulation!

Expect an even wider variety when a cloud is placed in the bottom row!

19 Next please

Introduction This activity, following on from 'Odds and evens', encourages pupils to find general terms for sequences, but does not use '*n*th term' notation. Sequences of numbers are related to patterns which give those numbers.

Materials OHP transparencies of sheets MI7 and MI8.

Possible content Rules for general terms for sequences.

Preamble The OHP transparency allows one 'square sequence' at a time to be shown. There are several places where pupils could discuss the issues involved in groups of two or three.

Present the first sequence of 'squares' on MI7.

"How many squares do you think will be in the next diagram, and why?"

Collect responses and reasons and discuss accordingly.

Referring to the same sequence, ask how many squares would be in the next diagram, ... and the next ... and the 10th ... and the 20th.

"How did you get these answers?"

Do not rush this stage. It is vital you establish the relationship between the position of the diagram and the number of squares it contains.

Tell the pupils you are thinking of a particular position in this sequence but you are not going to tell them which it is.

Represent this position by a cloud.
Write it like this:

1st	2nd	3rd	4th	...	(cloud)	...
1	4	9	16		?	

"Explain to me how you could work out how many squares this particular diagram has."

The correct response should be something along the lines of 'cloud times cloud', although the terminology used may vary.

This is an important step on the road to producing generalisations. At a later stage, the 'cloud' can be represented in the usual algebraic manner.

"How many squares are there in the 39th diagram?"

Collect and discuss responses.

"What position diagram would have 144 ... 81 ... 169 ... squares?"

Collect and discuss responses.

Present pupils with the second pattern of squares.

Follow through the same procedure as described above.

Complete the procedure to your satisfaction.

Tell pupils to think of and *then* draw a different arrangement of squares that will still produce the same sequence.

Pupils who are having problems could be shown the first two diagrams of other arrangements on the sheet.

The aim at this stage is to show that although the patterns formed may be different they are, in one sense, the same. The squares in the diagrams can be rearranged to form a much 'simpler' sequence, thereby enabling one to predict and/or describe further models more easily.

The following problem will enable you to demonstrate the value of this strategy, although some pupils may be able to generalise without recourse to it.

Show the pupils the following sequence of shapes on MI 8:

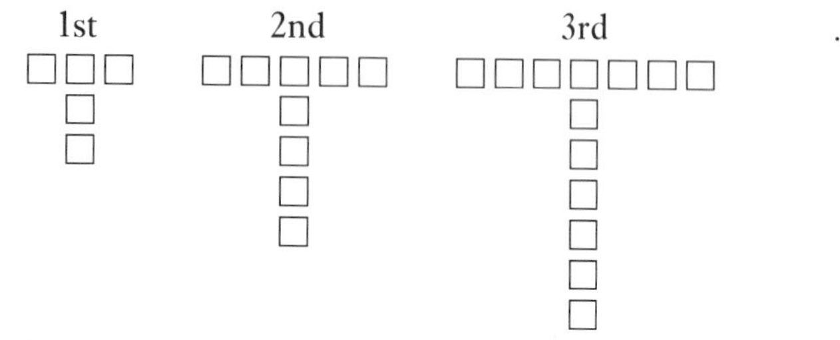

"Describe what the next diagram looks like ... and the next ... and the next. Explain why."

Make sure there is total agreement. Pupils will interpret the structure in different ways and these should be displayed, shared and discussed.

You may feel it is necessary to ask some pupils to draw the next few diagrams to check upon their understanding. These could be drawn on an OHP transparency or photocopies of MI 8.

Some of the likely interpretations are:

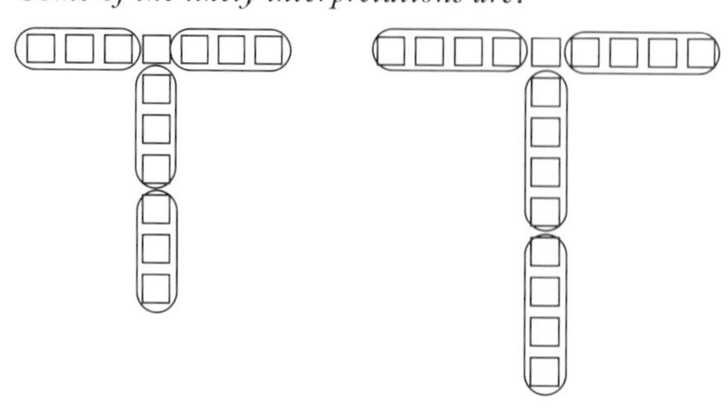

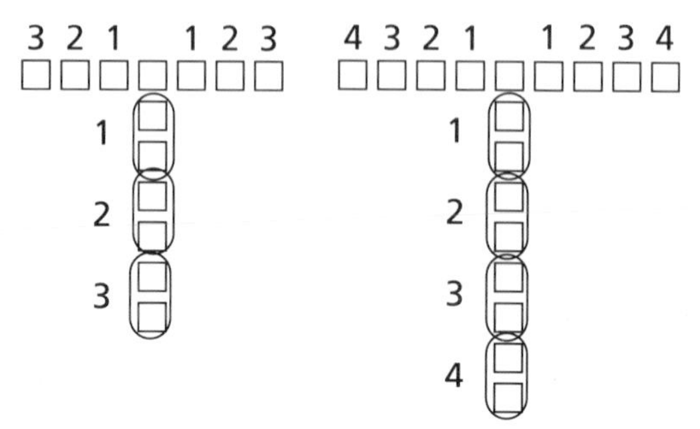

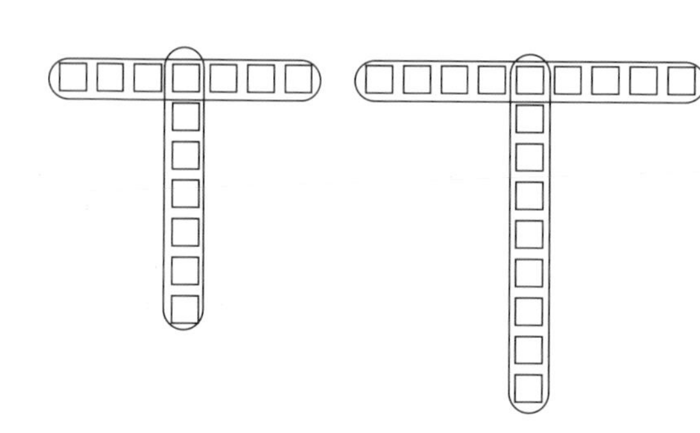

"Describe the 10th diagram. Describe the 35th diagram.
Describe the 'cloudth' diagram."

Do not rush this stage. Make sure there is total agreement. Again, different pupils will offer different descriptions depending on how they visualise and interpret the structure of the sequence and/or the individual diagrams.

"Now look carefully at how many squares there are in each diagram.
How many squares will there be in the 20th diagram?"

This might be a good opportunity for pupils to work in twos or threes so that they can share ideas, especially if you think the exercise would be less threatening! Each group must then come up with an agreed solution and reason.

For the above, the 20th diagram would give:

$20 + 20 + 20 + 20 + 1$ or $4 \times 20 + 1 = 81$

$20 + 20 + 20 \times 2 + 1 = 81$

$(20 + 20 + 1) + (20 + 20 + 1) - 1 = 81$

"How many squares would there be in the 'cloudth' model?"

The explanation should lead to a generalisation. The generalisations may be expressed differently although of course they could still be shown to be identical. You may wish to pursue this, depending on the understanding and ability of the pupils themselves.

The first interpretation gives an easy generalisation. The others are slightly more difficult.

$4 \times \text{cloud} + 1$

$\text{cloud} + \text{cloud} + 2 \times \text{cloud} + 1$

$(\text{cloud} + \text{cloud} + 1) \times 2 - 1$

If pupils (either certain groups or individuals) are having difficulty with the generalisation, then this would be an ideal opportunity to present them with the following rearrangements of the same number of squares in order to demonstrate the strategy mentioned earlier.

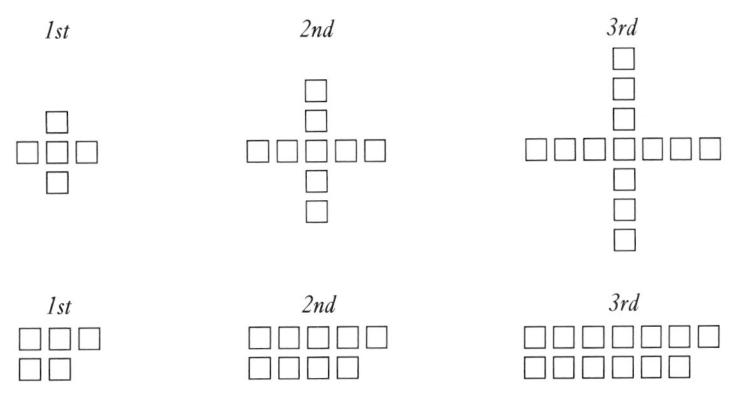

Pupils could be asked to make up two or more sets of patterns which give the same numerical sequence. They need not be patterns of squares. These patterns, which could be done as homework, could be used for display.

Square sequences 1

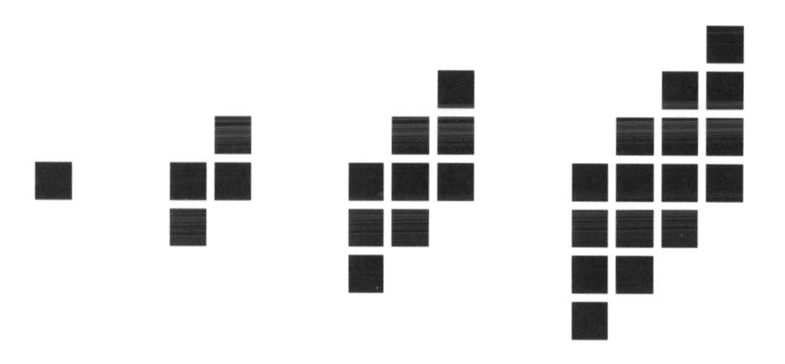

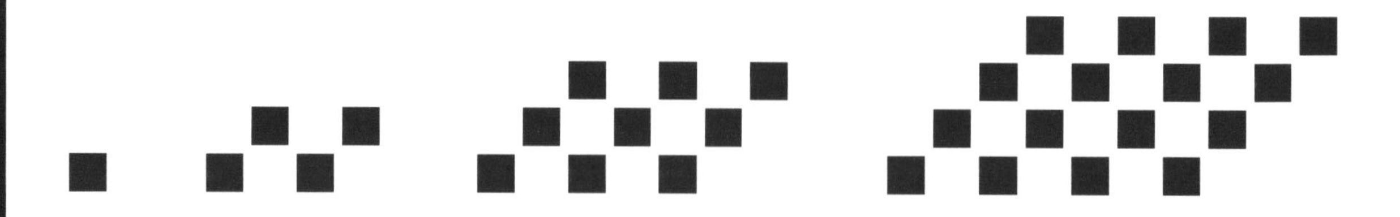

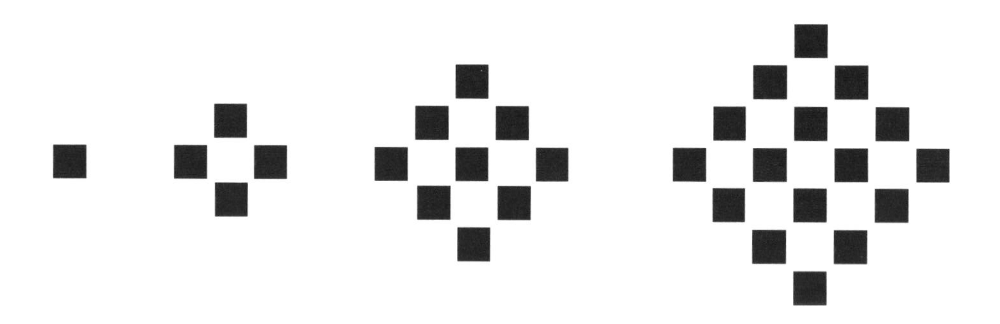

MI 8

Square sequences 2

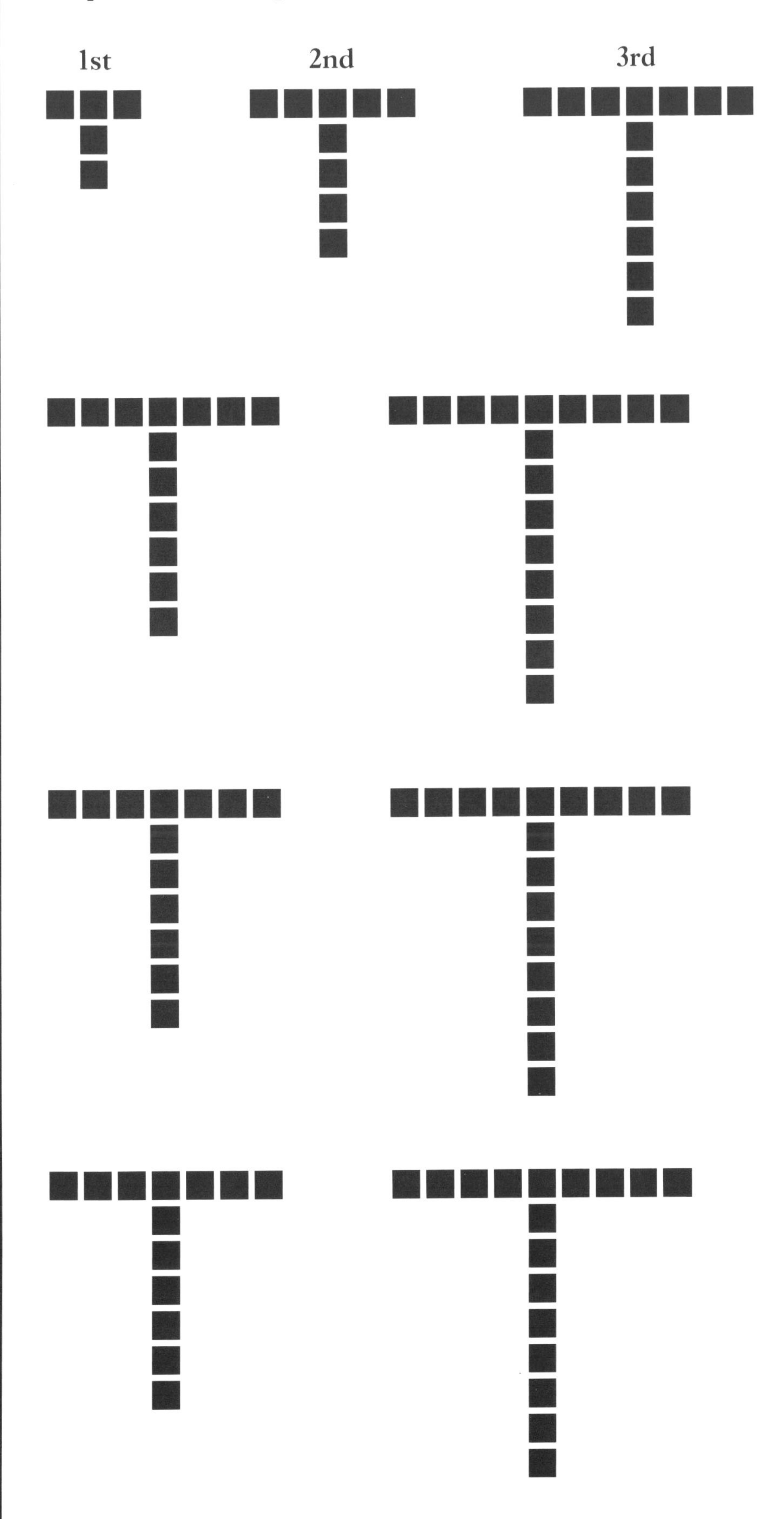

20 Tiles

Introduction This activity, following on from 'Next please', encourages pupils to visualise patterns and use this visualisation to generate the *n*th term for the pattern.

Materials Red and white chalk.

Possible content Rules for the *n*th term of a sequence.

Preamble By keeping the visual patterns simple it should be possible for pupils to keep the patterns in their minds and not draw all of them.

This activity is probably suitable only for more able pupils in years 7 and 8.

"I want you to imagine three red tiles in a row … with a white tile above each red tile … a white tile below each red tile … and a white tile at each end of the row of red tiles."

"How many white tiles are there?"

Ensure that pupils all agree with the correct answer and be prepared to repeat parts of the definition. Get one pupil to draw the pattern on the board after there is general agreement. Leave it there throughout the activity.

"Now imagine a diagram like this one but with a row of four red tiles instead of three. How many white tiles are there now?"

Again ensure that there is general agreement.

"How many white tiles would there be for a row of ten red tiles?"

Allow a number of responses to ensure all the group are thinking it through.

"How did you work it out?"

Establish a rule for the number of white tiles.

"How does your mental picture help you to explain how your rule works?"

To check that pupils have grasped the rule, get them to predict the number of white tiles for 40, 100, 39 red tiles.

"I am thinking of a number of red tiles. How would I work out the number of white tiles?"

Allow plenty of discussion to generate a general rule. Be prepared for apparently different rules and record them. Use the simple structure of the diagrams to convince pupils that the rules are equivalent. Symbolic representation could be introduced here using r to stand for the number of red tiles.

"Can we do it the other way round?
How many *red* tiles would there be for 12 *white* tiles?"

Make sure they realise they are given the number of white *tiles. Again ensure that they have the correct answer.*

"How did you work it out?"

Get clear explanations from some pupils. Again, be prepared for alternative methods.

"If I had w white tiles how would I work out the number of red tiles?"

Write down what pupils suggest as they say it: for example, for 'half of w minus two' write down $\frac{1}{2}w - 2$; for 'w minus two divided by two' write down $w - 2 \div 2$."

This might be the time to raise the issue of brackets or how to do the calculation on a calculator to ensure they understand that it should be $\frac{1}{2}(w - 2)$ or $(w - 2) \div 2$, respectively.

Often pupils are given diagrams from which they create a table of values. They are then asked to find a rule. On such occasions they should refer back to the original diagram to find and justify their rule.

On another occasion this exercise could be repeated with other 'simple' patterns, for example:

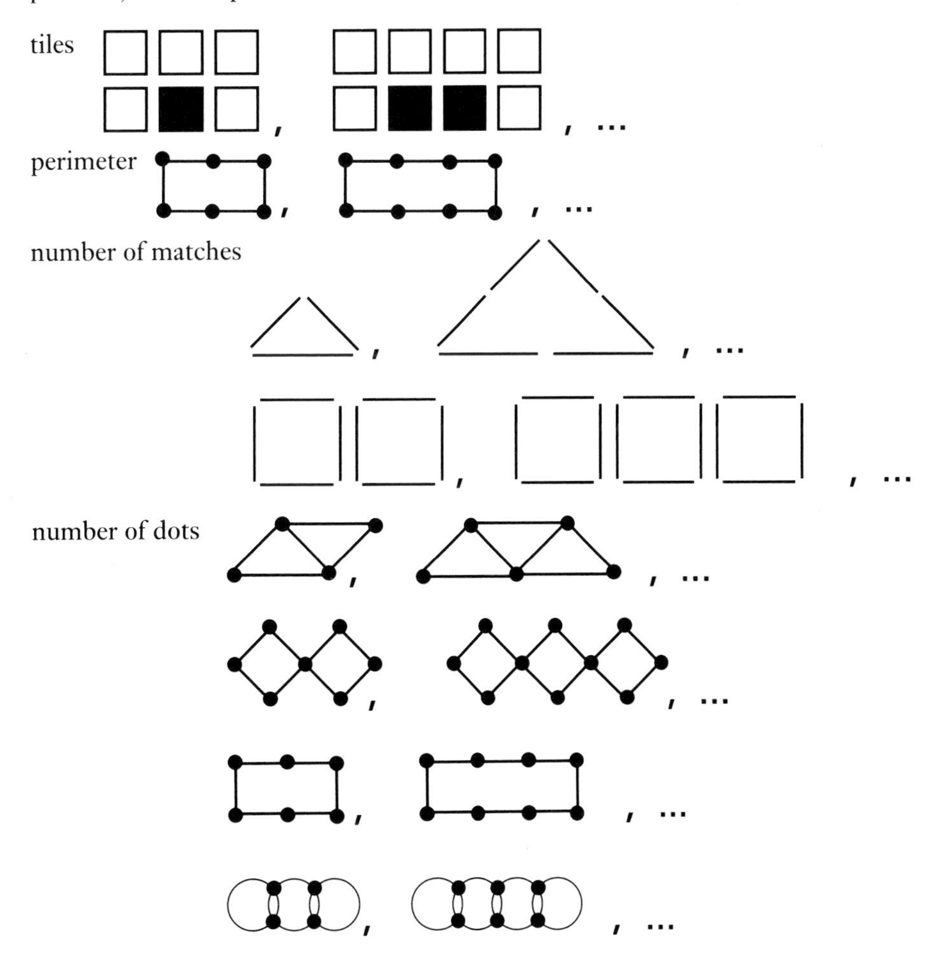

The activity could be done in small groups with one pupil playing the role of the teacher.

21 Chairs and squares

Introduction This activity encourages pupils to explain why apparently different algebraic expressions can be equivalent.

Materials Worksheet MI9. It may be useful to have this on an OHP transparency.

Possible content Algebraic manipulation; early ideas of proof.

Preamble This activity is best done after a situation arises naturally in class, for example during an investigation, where pupils arrive at different forms of the same expression, or after 'Tiles'.

The activity is suitable for more able pupils in year 8.

"Tables and chairs are arranged like this."

Show these arrangements of tables and chairs on the board or OHP:

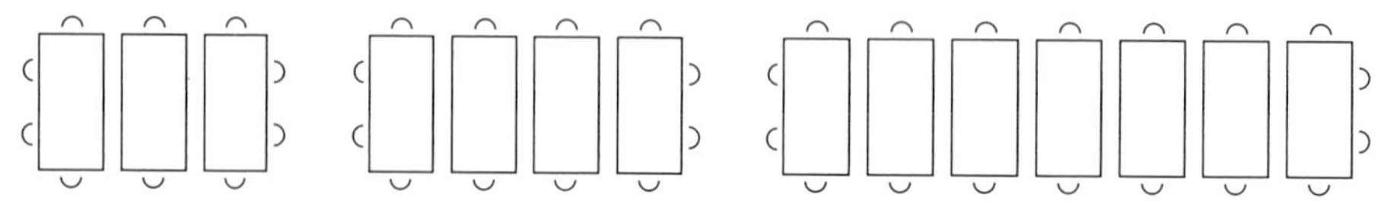

"Suppose we wish to find a connection between the number of tables and the number of chairs. Remember, your rule must work for any number of tables. We can look at the diagrams in different ways."

Show the diagram on MI9.

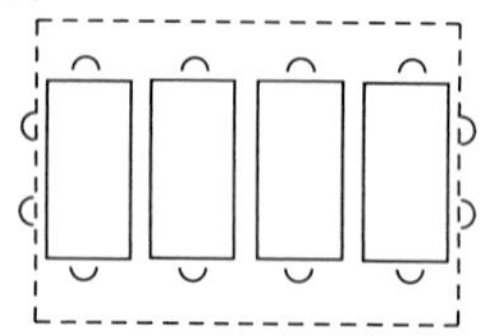

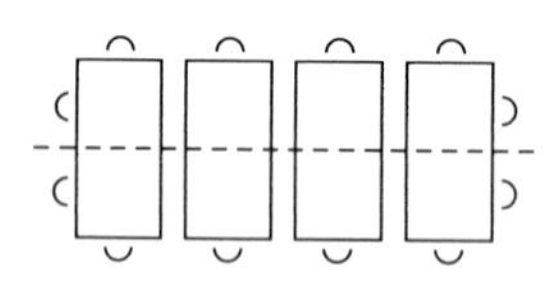

"What rule does each diagram suggest for getting the number of chairs from the number of tables?"

Be prepared for many different suggestions (including 'sawing the table in half!' for the second diagram) but ensure that $c = t \times 2 + 4$ *and* $c = (t + 2) \times 2$ *appear either in symbols or words.*

"We have looked at the diagram in different ways.
The tables and chairs are the same in each so the rules must agree.
The formulas look different but they must be different ways of expressing the same rule."

"Imagine a pond which is 1 metre square.
Surround it with a path of paving stones each 1 metre square.
How many paving stones does it need?"

If you get the answer '4' tell them you need to be able to walk round the pond without jumping, so you want the corners paved.

"Imagine a pond 2 metres square surrounded by a path. How many paving stones will I need?"

"What about a pond 3 metres square?"

Work through the examples fairly quickly and generate a table of results on the board.

"What rule can I use to get the number of paving stones from the length, l, of the side of the pond?"

Encourage and accept different forms of the rule (accepting the incorrect $4(l+2)$ *at this stage). Make sure that you have both* $4l + 4$ *and* $(l+1) \times 4$.

"We are now going to draw some diagrams to help justify these rules. We will draw only one size of pond, but remember your explanation of the rule must work for any size."

Show the diagram for a pond with side of length 3.

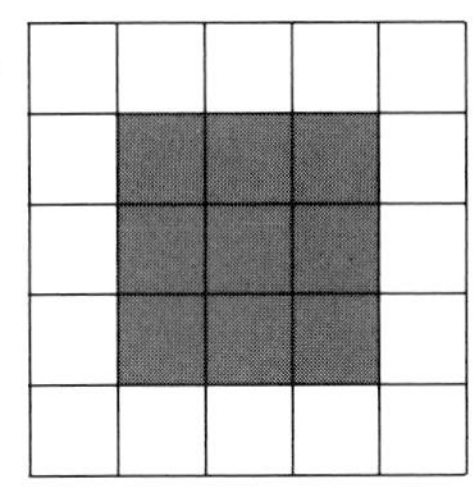

"Use two copies of this diagram on MI9. Try and use one diagram to show why $4l + 4$ must be a rule for the number of paving stones and the other to show why $(l+1) \times 4$ must be a rule."

If pupils get stuck or are not sure what to do, then give them the idea for $4l + 4$ *and get them to show why* $(l+1) \times 4$ *must be a rule.*

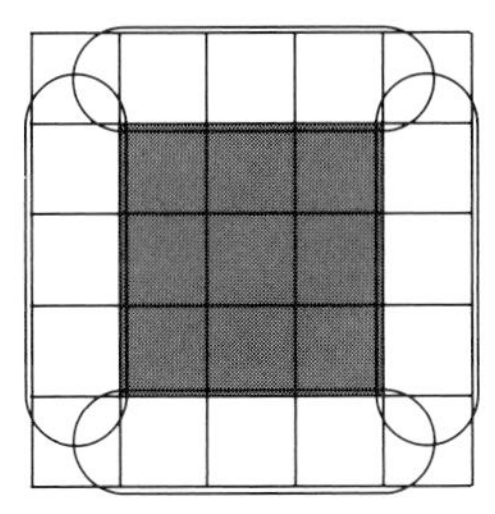

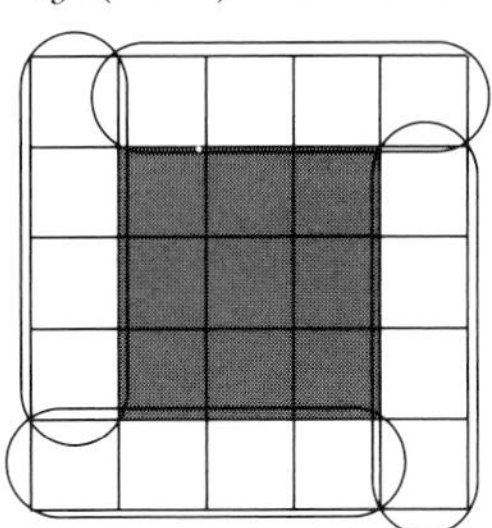

"The length of the side of the large square is $l + 2$.
On another copy of the diagram show whether or not $4(l+2)$ is a rule for the number of paving stones."

Give pupils time to attempt this problem.
Discuss the fact that the corner squares are counted twice, so the rule should be $4(l+2) - 4$.

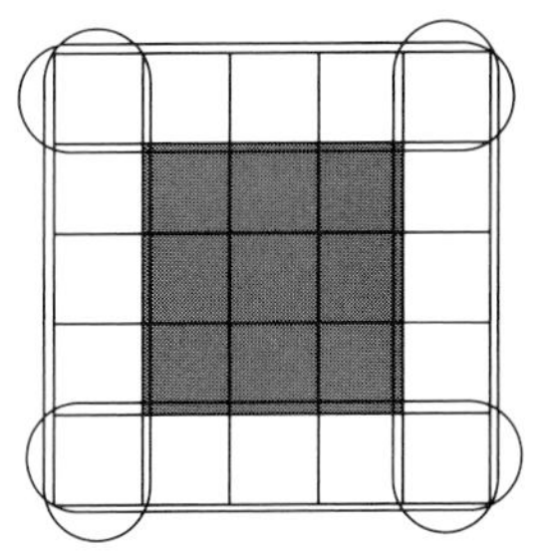

"What rules are suggested by these ways of looking at the diagram?"

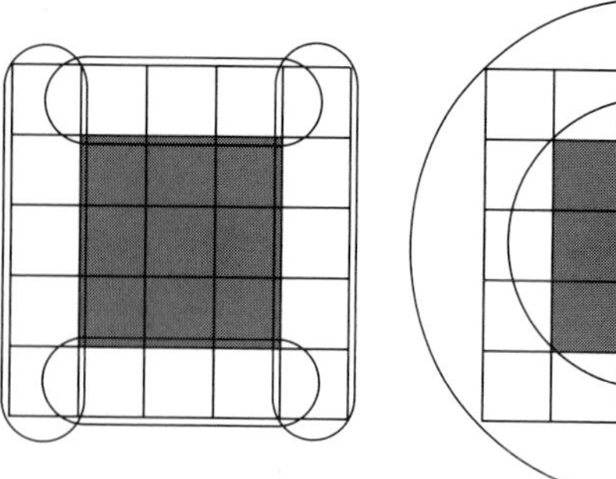

Establish $2(l + 2) + 2l$ *and* $(l + 2)^2 - l^2$.

"Again all these rules must be equivalent as the number of paving stones is the same in each case."

"The diagrams show that these rules are equivalent for any value of l, not just $l = 3$."

Draw or show this diagram from MI9.

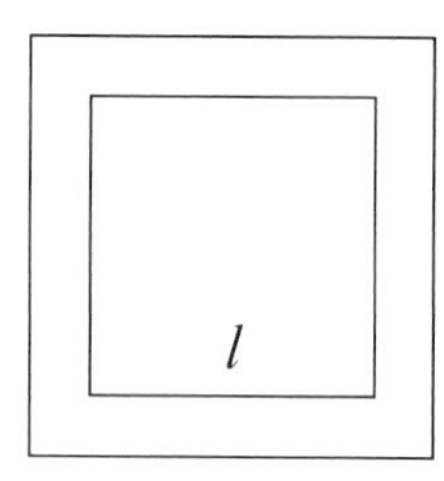

"Imagine a square lawn surrounded by paving stones, in the same way that the pond was surrounded. The paving stones are 1 metre square."

"What is the length of the side of the large square?"

Ensure all understand that it is $l + 2$.

"What is the total area of the lawn and paving stones?"

Focus on $(l + 2)^2$ *as the area, but do not reject other suggestions.*

"If we look at the garden in these ways, what rules for finding the total area does each one suggest?"

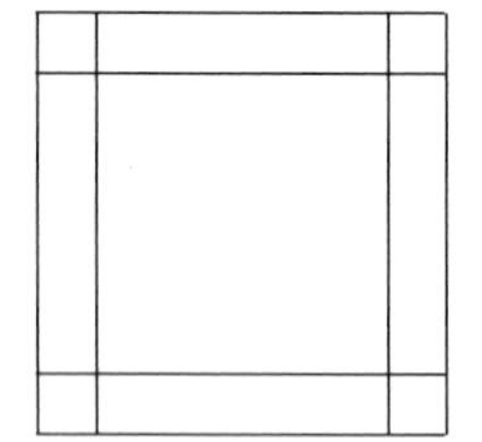

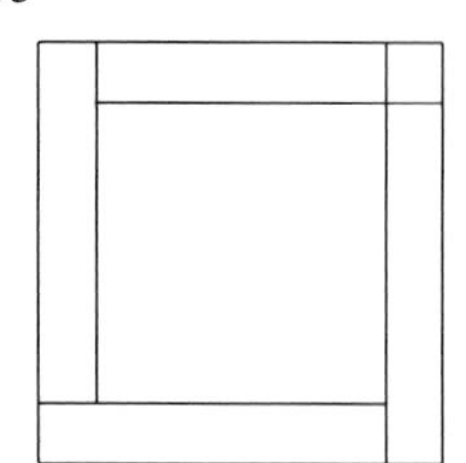

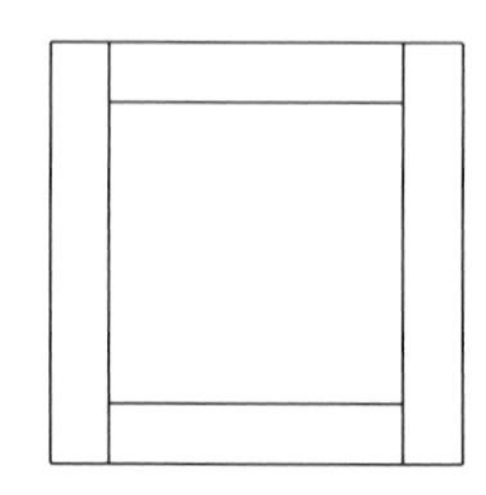

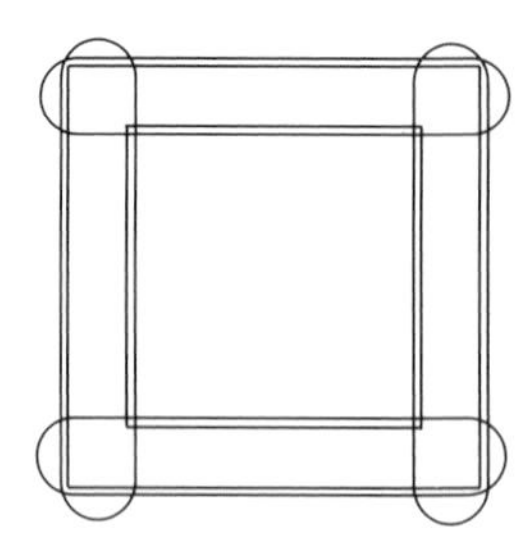

Establish $l^2 + 4l + 4$, $l^2 + 4(l + 1)$, $l^2 + 2l + 2(l + 2)$, $l^2 + 4(l + 2) - 4$, *respectively.*

"These forms of the rule look very different but must be the same, as the area is the same in each case."

"If we put $l = 3$ in each form of the rule, we get the same answer, 25. This does not **prove** that each form of the rule will give the same result for other values of l."

"By thinking about the diagrams, we have seen that the different versions of the rule are equivalent for all values of l."

Name ....

Chairs and squares

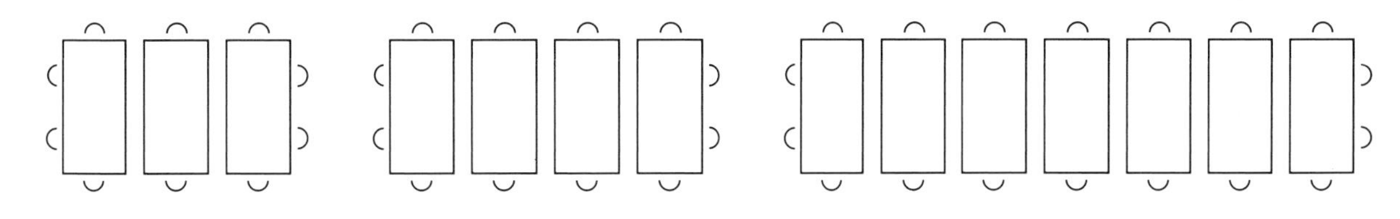

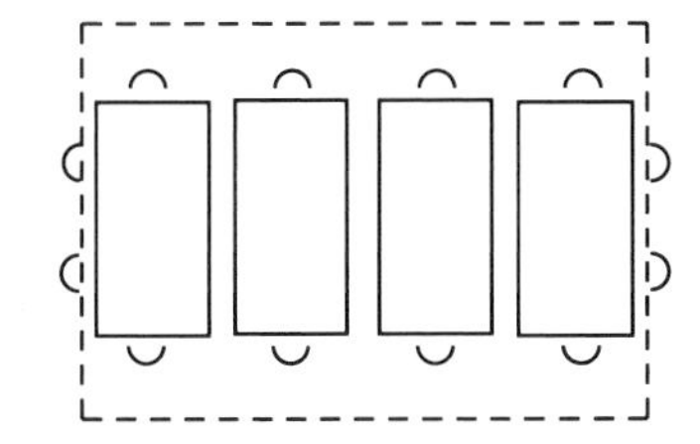

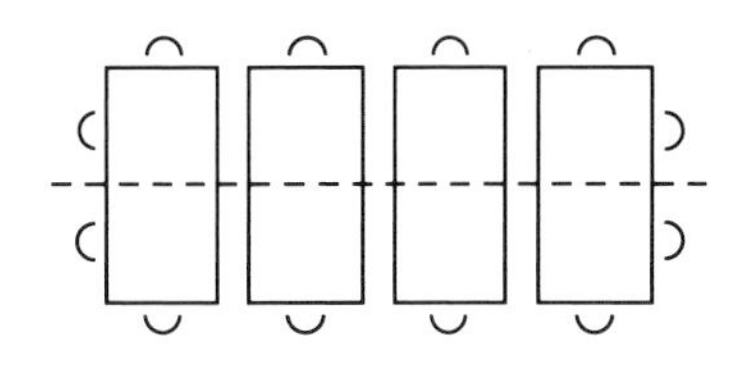

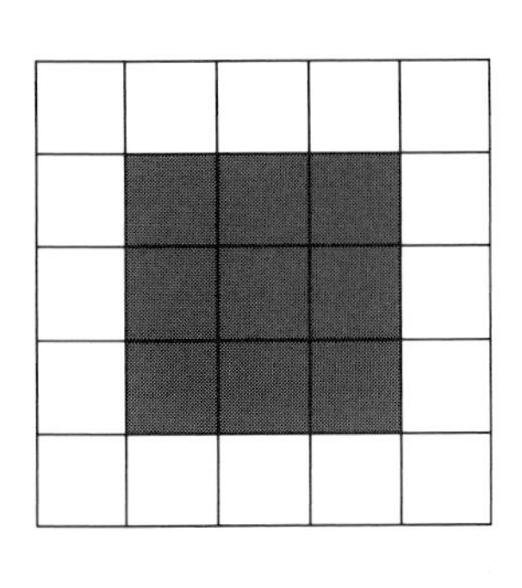

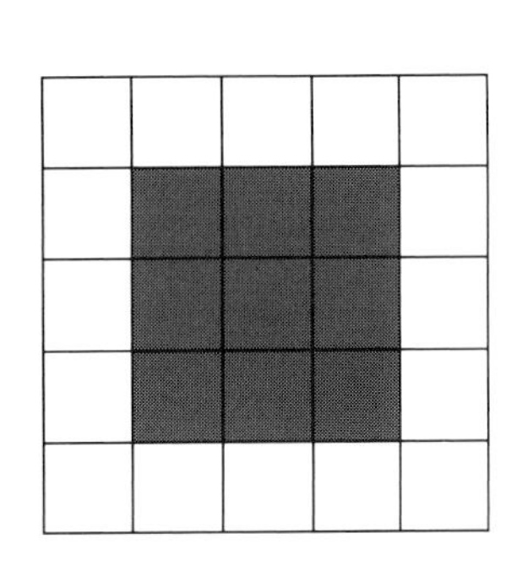

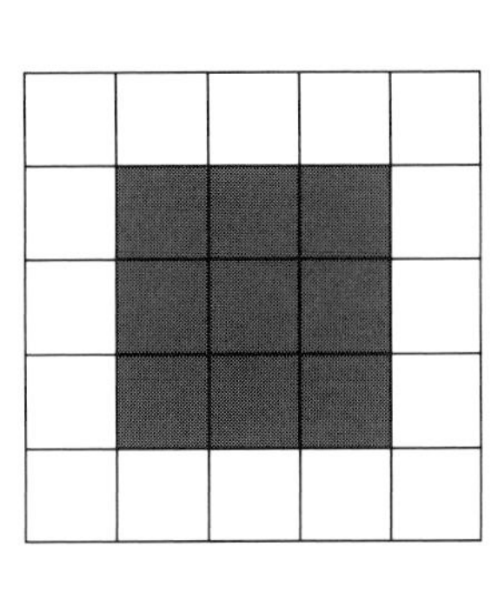

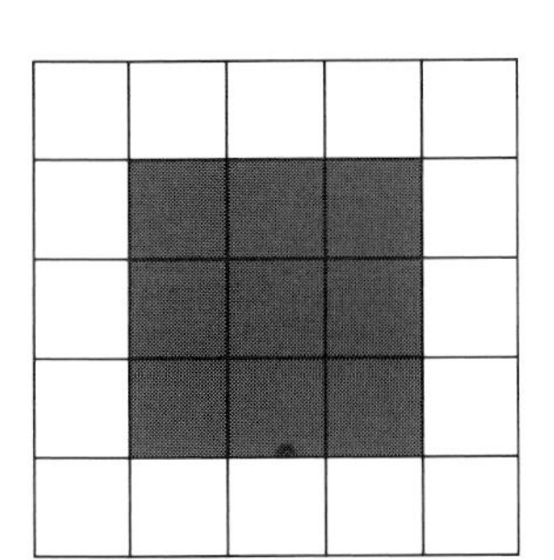

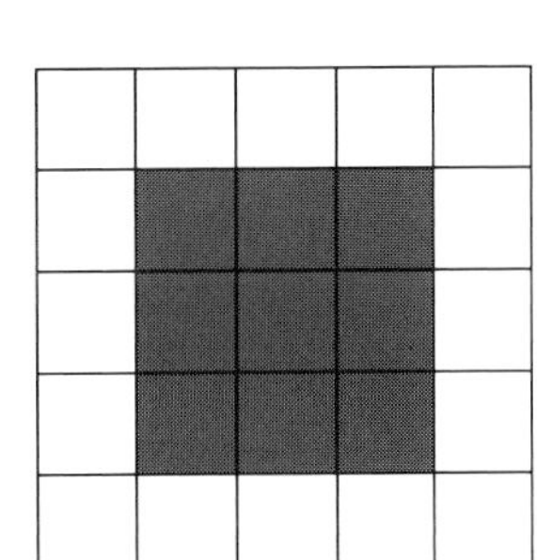

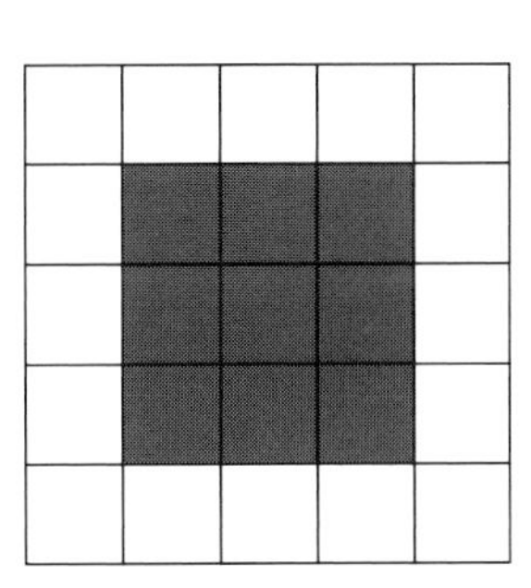

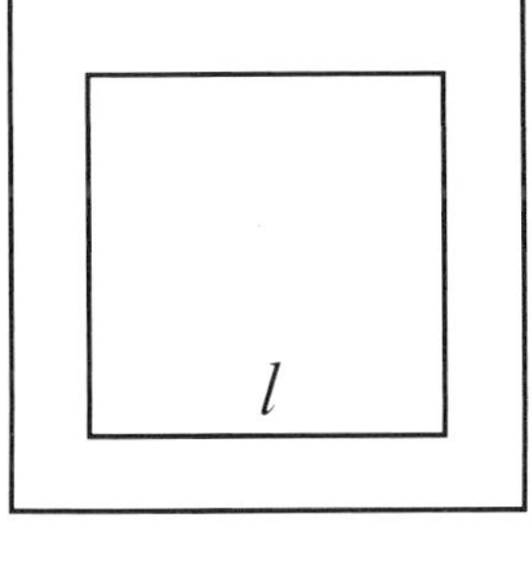

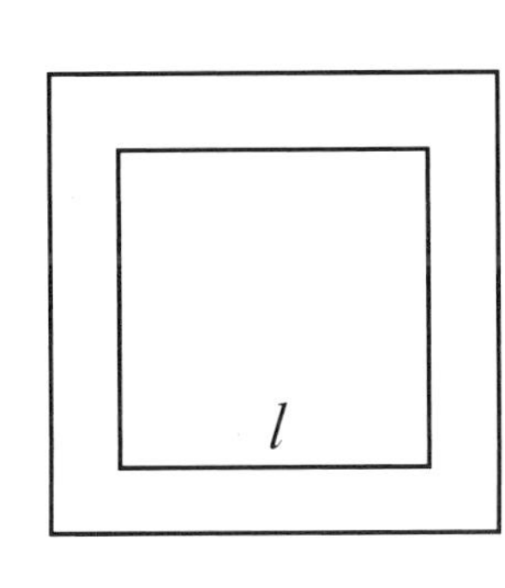

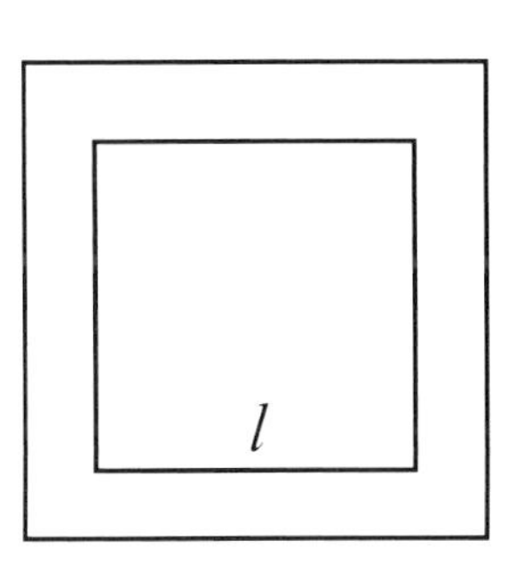

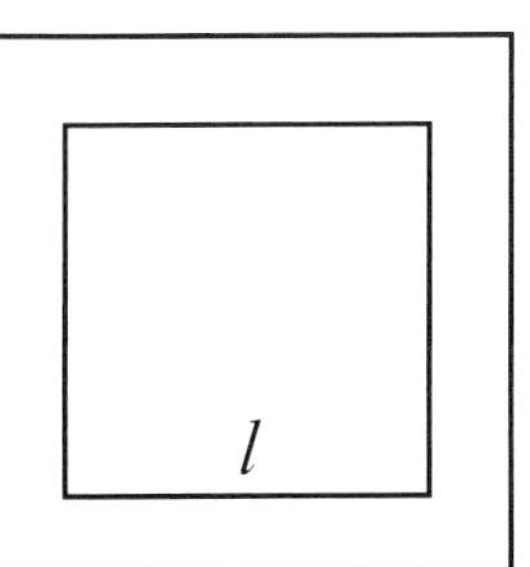

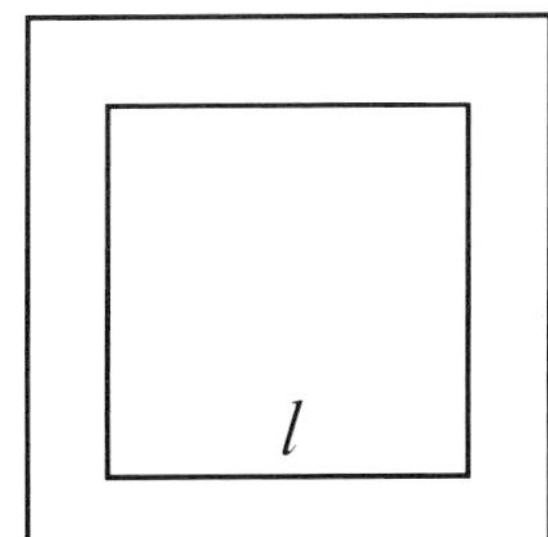

l

22 Watch the angle

Introduction The aim of this activity is to help pupils visualise angles of a given size, and to estimate the size of a given angle, using a clockface to provide basic 'angle units'.

Materials A clockface, with either no hands or geared hands; worksheet MI 10.

Possible content Visualising, comparing and estimating angles.

Preamble Pupils will need to know that the angle on a straight line is 180° and that a complete turn is 360°.

Although this activity is based on a clock, it is not to do with time.

During the course of the activity other interesting mathematical problems, some based on time, will arise. These should not be allowed to distract pupils from the main thrust of the work.

"When do the hands of a clock point in exactly the same direction?"

Most pupils will suggest 12 o'clock.

Finding the other ten times is an interesting problem, but should not be pursued here. If times like 'half past six' are suggested, establish, preferably by discussion, that the hands are not 'exactly' together as the hour-hand has moved on.

"What is the angle between the hands then (for example at 12 o'clock)?"

This question is asked to indicate that the activity is about angle, and not time.
You may find that some pupils have difficulty with angles of size 0°.

"When do the two hands point in exactly opposite directions?"

Most pupils will suggest 6 o'clock.

Again, do not try to find the other 10 times. If it is raised, establish that at times like 'a quarter to three' the hands are not exactly opposite each other.

"What is the angle between the hands then (for example at 6 o'clock)?"

Establish 180°.

"So we can get 0° and 180°. Can we get exactly 90°?
Is there a time when the angle between the hands is exactly 90°?"

You should find that pupils will quickly suggest 3 o'clock and 9 o'clock. If you only get one of these, then prompt for the other.

You may also get suggestions such as five past four or a quarter past six. If you have not had the incorrect results for 0° or 180°, then you should raise the question at this point. Discuss with the class whether the angle is more than 90° (for example at five past four) or less than 90° (for example at half past three) and how we know this. Do not try to find what the angle actually is.

Again, do not pursue the task of finding all the times when the angle is exactly 90°.

"What other simple angles could we look for?"

Accept and record suggestions. Discuss what is meant by 'simple' in this context.

Home in on the basic unit of angle on the clockface of 360° ÷ 12 = 30°.

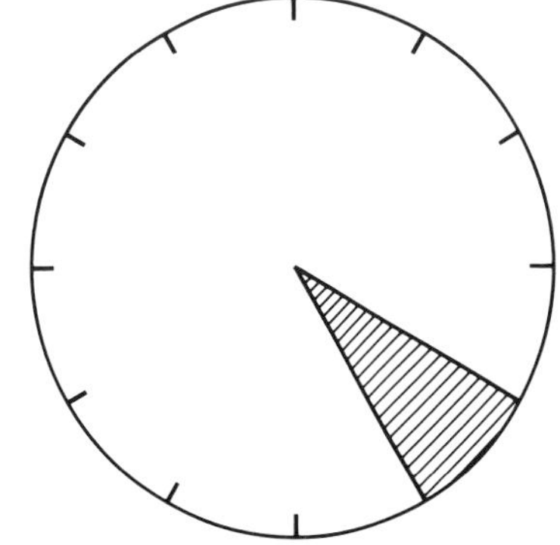

"When do we get an angle of exactly 30°?"

Establish 1 o'clock and 11 o'clock.

Use this to reinforce the 360° ÷ 12 *idea if necessary.*

"So what is the angle between the hands at 2 o'clock?"

Establish that 2 × 30° *can be used.*

Ask sufficient questions of this nature to ensure that the 30° unit is accepted as a standard mental image.

"Now can you work out the angle between the hands at half past three?"

Some prompting may be needed to get pupils to realise that the hour hand has moved on through 'half a unit'.

You may wish to ask further questions like this to consolidate the ideas of using the clockface to help picture angles of 15°, 30°, 90°.

Depending on the aptitude of the class, you could ask questions like:

(Given the time, what is the angle) – "What is the angle at 7 o'clock?"

(Given the angle, what is the time) – "When would we get an angle of 120°?"

(Questions involving part units) – "What is the angle at 2:30?"

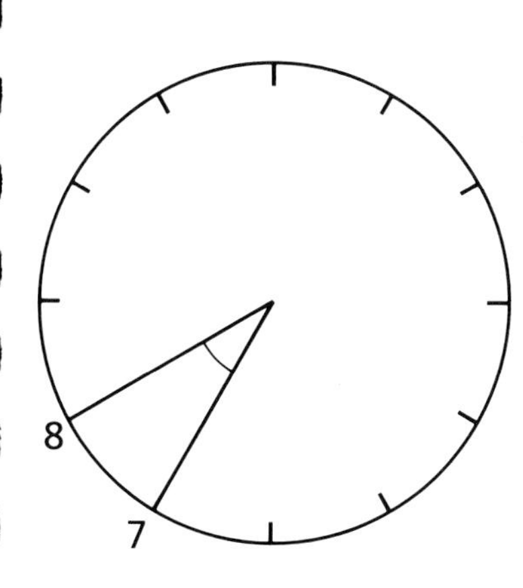

The purpose of such questions is to make pupils aware that the angle subtended at the centre by any two adjacent hour marks (for example, 7 and 8 o'clock) is 30°, even though there is no time at which the hands of a clock will actually point exactly at them.

This can help in visualising the sizes of angles in different orientations.

If appropriate, you could also draw out that the angle between the hands at 4 o'clock or 8 o'clock represents a third of 360°.

Now issue worksheet MI10.

"Think of the clockface to help you answer these questions."

"Using only a ruler and a pencil, I want you to draw some angles as accurately as you can."

Use a selection of the angles listed.

"Draw an angle of: 30°, 60°, 90°, 120°."

These are all based on the 30° unit.

"Draw an angle of: 75°, 45°, 135°, 15°."

These are based on the 'half unit'.

If appropriate, you could ask pupils to draw an angle of: 10°, 50°, 20°, 170°, 100°.

These are based on 'a third of a unit'.

"Some angles have been drawn on the worksheet. Estimate their sizes."

Possible extensions

- Find *all* the times when the angle between the hands is of a certain size, for example 180°.
- Ask more searching questions about angles on the clockface, such as:

  "What is the angle between the hands at 5:30? 11:30? 8:30?"

  "At what times is the angle equal to 15°?"

  "We know that at 3 o'clock the angle is exactly 90°, but when is it exactly 45°?"

  "Can we get an angle of exactly 75°? 10°? 20°? 40°? 50°?"

  "What is the angle at 1:40?"

  "What is the first time after 4 o'clock when the angle is 10°?"

Name ....

Watch the angle

Draw the angles in this space:

Estimate the sizes of these angles. Write your answers on the diagrams.

23 Clock polygons

Introduction This activity uses a mental image of a clockface to construct a number of polygons mentally.

Materials It would be useful to have a clockface or a copy of the top of sheet MI 11 available for pupils who find the activity too difficult to do mentally. For 'Clockface bingo', you need a bingo card for each pupil and a set of caller cards (sheets MI 11, MI 12A and MI 12B); counters or small pieces of card to cover the shapes.

Possible content Two-dimensional shapes; congruence.

Preamble The clockface has the advantage of being familiar and providing an easy way to refer to the points around the edge of a circle. Even if pupils use the clock on sheet MI 11 throughout, a lot of mental imagery will be required if they use no other aid.

Pupils need to be very familiar with the positions of numbers on a clockface and to know the names of the common quadrilaterals. The extensions explore the language of triangles. The game 'Clockface bingo' gives practice in identifying the polygons.

Some pupils may use a mixture of mental images and numerical differences to answer the questions. There is nothing wrong with this and it could form the basis of some informal proofs.

"Close your eyes and keep them shut even when giving answers.
Imagine a clockface. Make sure that it has a mark for each hour.
Which hour mark is at the top?
Which hour mark is at the bottom?
Now look at the hour mark furthest to the right. Which is it?
And which is furthest to the left?"

The opening questions are designed to establish the way language will be used and to ensure that pupils have a common mental image.

"Imagine the shape you get if you join 9 to 12 … 12 to 3 … 3 to 6 … 6 to 9."

Pause in between each pair to allow pupils time to think and produce their mental image.

"What would you call it? Has it a special name?"

Expect the answer 'a diamond'. Establish that it is actually a square, by asking questions such as the following:

"What do you mean by 'a diamond'?"

"How many sides has it got?"

"What lines of symmetry does it have?"

"Which side is longest?"

"Which angle is biggest?"

"Has the shape got another name?"

If pupils still have difficulty, you can show a clockface and rotate it or use the diagram of a clock on sheet MI11.

"What shape do you get if you join 10 to 1, 1 to 4, 4 to 7 and 7 to 10?"

Establish that this too is a square.

"Can you see any other squares on your clockface?"

Allow time for pupils to think. If necessary, you can prompt them by asking if there is a square which starts at 2.

"How many different squares are there?"

Establish that there are only three.

"Are they really different?
Or are they all congruent, that is the same size?"

Establish that they are all congruent.

"In your head, imagine the shape you get if you join 10 to 2, 2 to 4, 4 to 8 and 8 to 10."

Pause to allow pupils time to construct their image.

"How many sides has it got? Does it have a special name?"

Once most of the group have identified it as a rectangle, move on.

"How many different rectangles can you see on your clockface?"

Pupils may well stick to ones with a horizontal base and imagine that 5 – 7 – 11 – 1 *and* 4 – 8 – 10 – 2 *are different.*

Establish that there are just three types (pupils will forget that a square is a rectangle).

"Can you see a rhombus which isn't a square?"

"Can you find a parallelogram which isn't a rectangle?"

"Can you see a trapezium? How many different trapeziums are there?"

These are quite hard. It may be a good time to produce a clockface and to get pupils to open their eyes.

Possible extensions

- Can you find an equilateral triangle? How many of them are there?
- Can you find any isosceles triangles? How many types are there?
- Can you find a right-angled triangle? How many types are there? Do they all go through the centre of the clock?
- How many different obtuse-angled triangles are there?
- How many different acute-angled triangles are there?
- Can you find a regular hexagon?
- Which other regular polygons can you find?

■ Clockface bingo

This is an activity for a large group.

Give out the bingo cards. (The arrow in the top left-hand corner shows which way is up.) The caller cards should be cut up and picked at random by the teacher or a pupil who calls out the numbers. Allowing a limited length of time between cards should encourage pupils to visualise the shape and find a match on their card. Otherwise they may check the number sequence against each item in turn on their card, which will not be as helpful with identifying polygons. Players cover up the appropriate shape on their card. The first person to cover all their shapes wins. The shapes can be checked against the cards which have been used.

Clock polygons

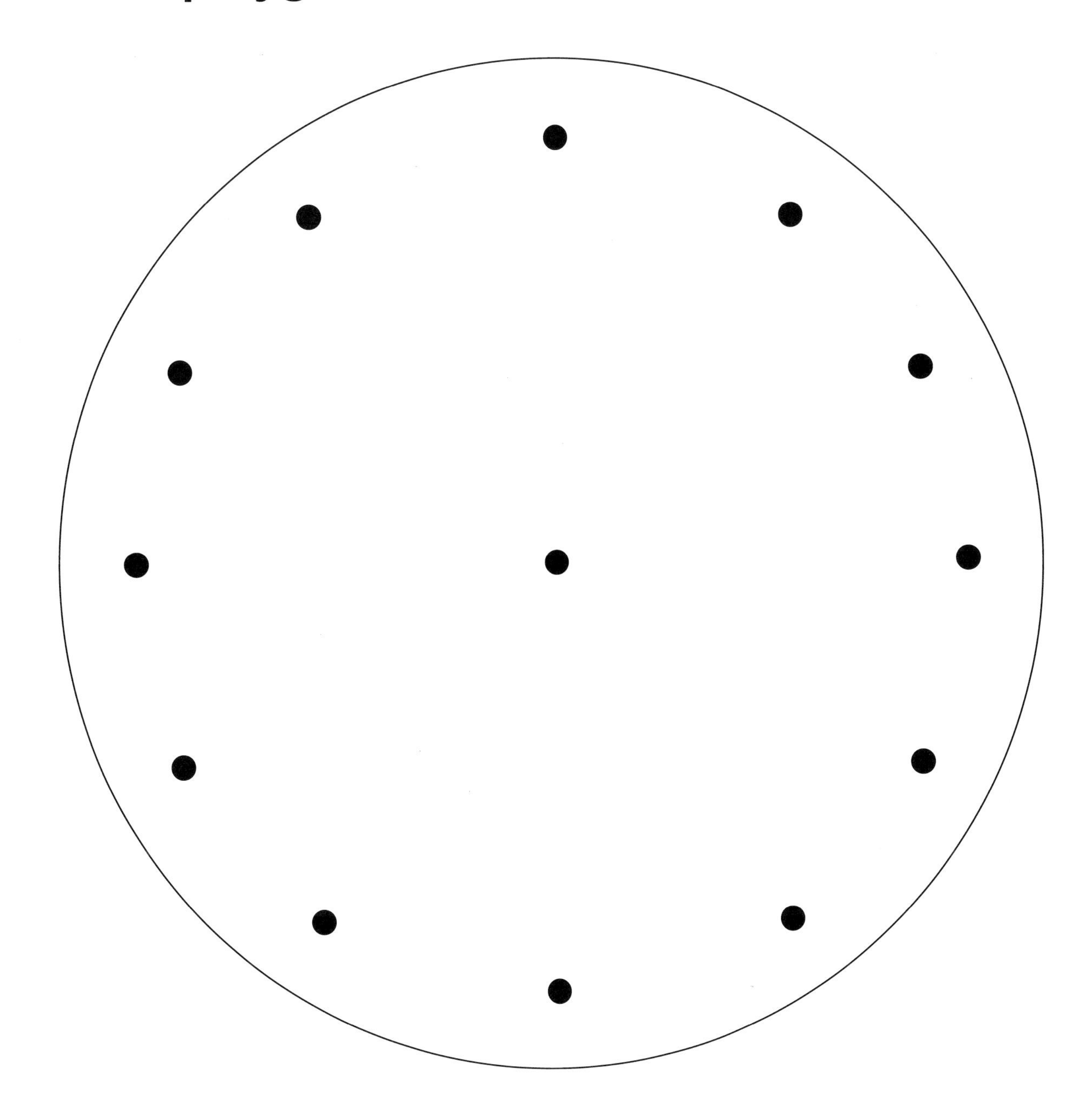

Clockface bingo caller cards

12, 4, 8	10, 2, 6	9, 1, 5
3, 7, 11	12, 6, 9	9, 12, 3
12, 3, 6	3, 6, 9	11, 1, 5, 7
10, 2, 4, 8	10, 11, 4, 5	1, 2, 7, 8
12, 3, 6, 9	1, 4, 7, 10	2, 5, 8, 11
1, 3, 5, 7, 9, 11	2, 4, 6, 8, 10, 12	11, 1, 4, 8
2, 4, 7, 11	5, 7, 10, 2	8, 10, 1, 5

Clockface bingo cards

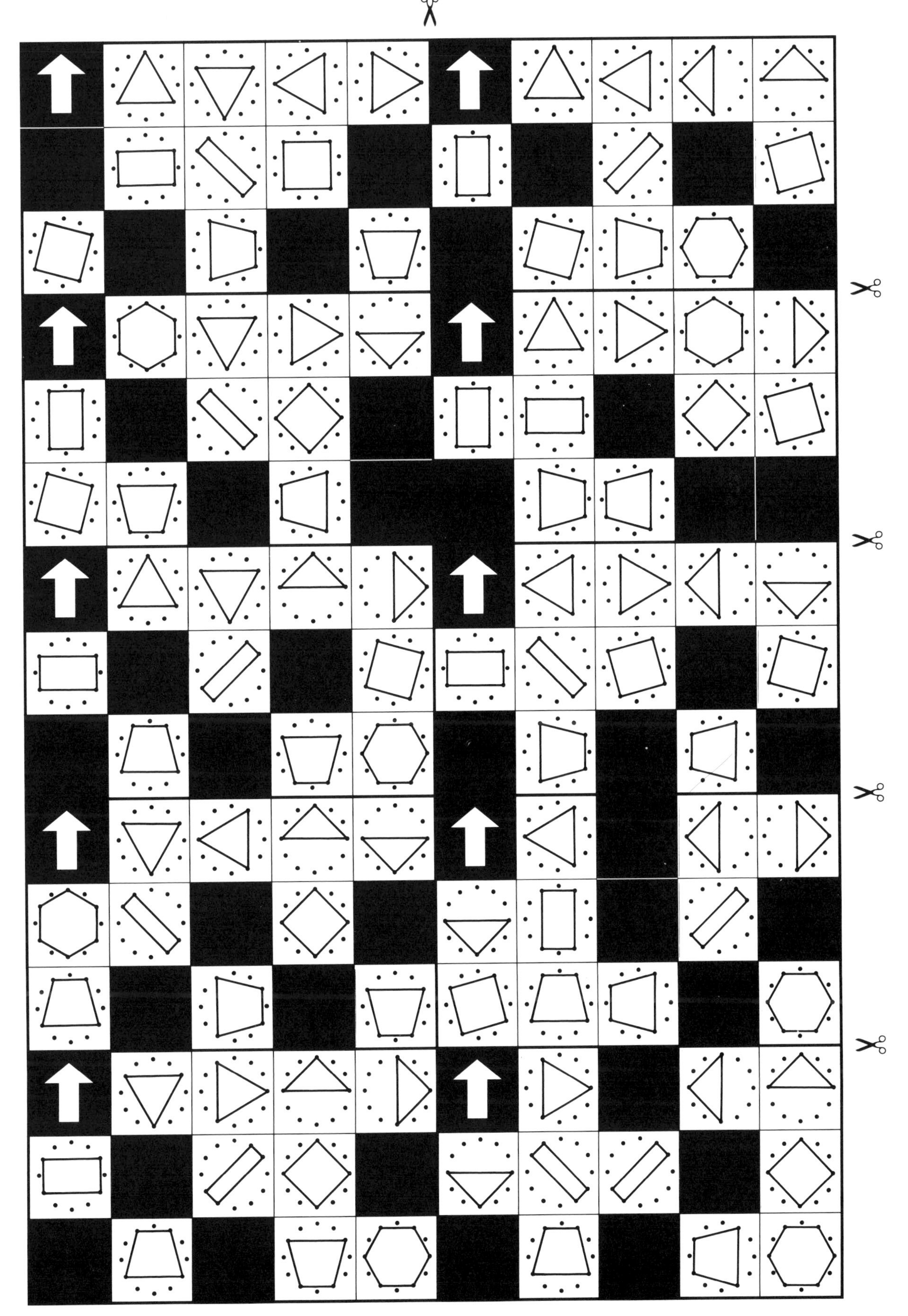

24 Roundabout

Introduction In this activity, pupils order a set of views of a familiar object.

Materials Two sets of cards: six drawings of a television on sheet MI 13 and eight drawings of a van on MI 14.

Possible content Using two-dimensional representations of three-dimensional objects; recognising orientation.

Preamble Pupils do not find this type of activity as easy as it appears. They need practice. The activity is probably best done in pairs, to encourage discussion.

Give the pupils one of the sets of cards. Tell them that they have to put the pictures in the order they would see them as they walked round the television or van.

For more able pupils, you could also specify clockwise or anti-clockwise.

A harder version of the activity is to leave the sheets uncut and ask pupils to list the order in which the line drawings should be viewed.

This skill should be given as much practice as possible. We have provided two sets of drawings as a starting point, but the activity becomes much more attractive to pupils if you can produce your own sets of photographs of local objects. You should use everyday objects which are familiar to pupils, such as a chair, a shoe, a shed or a person. When taking the photographs you should remove as much background as possible, so that the object concerned is the main visual focus.

Perhaps you could get the Art department or the resources technician to take the photographs. A disposable camera will give 24 photos – three sets of eight – at a reasonable cost. Another possibility is to ask the school photographer to take a series of pictures of your head.

MI 13 Roundabout 1

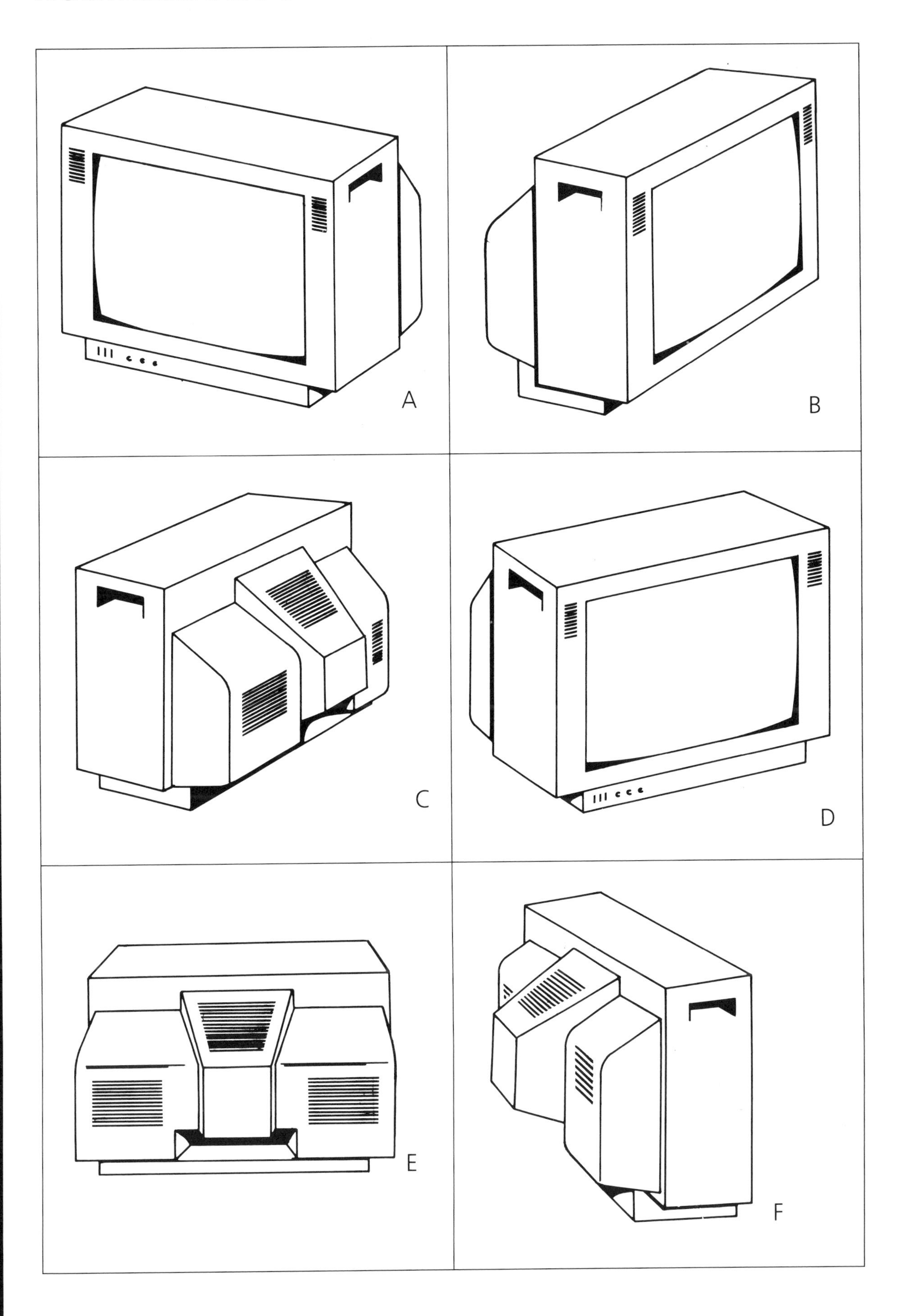

Roundabout 2

25 Everyday objects

Introduction The aim of this activity is to guess what object is being described mathematically and to describe objects using mathematical language.

Possible content Recognising the names of shapes; estimating lengths.

Preamble It would be a good idea to try this activity soon after doing 'Maths around us'. This activity comes in three parts. In the first, pupils are given descriptions of objects and they must decide what is being described. In the second, pupils describe given objects just using mathematical language, and in the third they think of their own objects to describe. Part 2 or 3 could be done as homework with pupils trying out descriptions on a partner in the next lesson. They can be written up as a wall display for other classes to guess. It is important to emphasise that the descriptions are only approximate.

Guessing the object

Each of these descriptions can be read out a couple of times. Ask pupils to put up their hands when they know what the object is. The clue in brackets could be given if very few pupils have put up their hands.

1 This object is a sphere on top of a long, thin cylinder. The sphere is about 30 cm in diameter, and the cylinder is about 3 m tall and about 10 cm in diameter. (The cylinder has black and white bands.)

2 This is a hexagonal prism which is about 18 cm long and half a centimetre across and it is usually shaped to have a cone at one end. (You sharpen it.)

3 This object is made from a thin cuboid standing on four thin cylinders. The cuboid is about one metre by one and a half metres by 2 cm. The cylinders are about a metre high and 5 cm in diameter. (It may be made of wood.)

4 This is a long, thin cylinder about 30 cm long and half a centimetre or less in diameter. The cylinder has a cone on one end and a disc about 1 cm in diameter on the other. (They are usually in pairs.)

5 This object is a sphere which is covered completely by hexagons and pentagons. It is about 30 cm in diameter. (It is usually made of leather.)

6 This is a disc with a cylinder attached to the rim of the disc so that it is an extension of the diameter. The disc is about 20 cm in diameter, and the cylinder is about 3 cm in diameter and 10 cm long. (It is used to hit a small white sphere.)

7 This object is a long thin cylinder about 3 m tall and 10 cm in diameter with an equilateral triangle fixed to the top with a corner pointing downwards. (It is in *The Highway Code*.)

Answers: 1 a Belisha beacon; **2** a pencil; **3** a table; **4** a knitting needle; **5** a football; **6** a table tennis bat; **7** a 'give way' sign.

Describing objects

Ask pupils to describe an object using only mathematical language and avoiding any mention of colour, material it is made from or use. Some examples they could try are a matchbox, a PrittStick, a washing-up bottle, a pillar-box, a book, a metre rule, a clock, a cotton reel, nuts and bolts, washers.

Inventing their own

Pupils can be asked to write out a description of an object of their own choosing and to try out this description on a friend.

26 Back-to-back

Introduction The aim is to help pupils appreciate that information has to be communicated accurately and precisely; to develop their ability to transmit and receive information; to enhance their powers of mental imagery and to develop a mathematical vocabulary and the use of mathematical language.

Materials Paper for pupils' diagrams, enough copies of the eight 'Back-to-back' cards on sheet MI 15 for the whole class.

Possible content Mathematical language involving shape and angle.

Preamble It is important to keep the first few diagrams very simple to establish the need for precision. As measuring instruments will not be used, it is the ratio of the sizes of the various shapes that is important not their absolute size. This is an activity that can be returned to a number of times using more complex shapes.

Draw this on a piece of card.
Do not show it to the pupils.

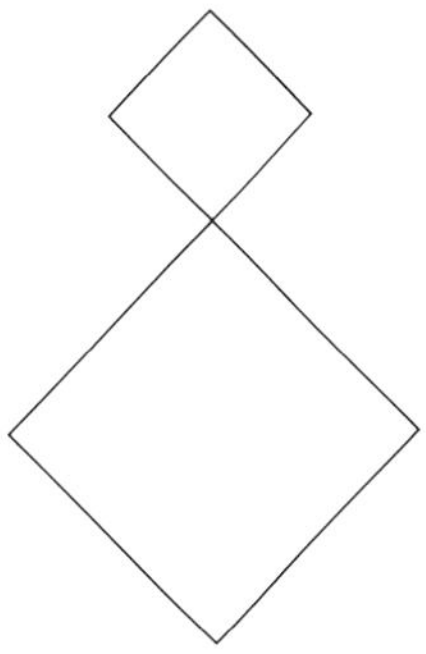

"I have a diagram which I am going to tell you how to draw. I want you to draw it as accurately as you can. I will only give the instructions once and I won't answer any questions."

"Draw a square."

Give pupils time to draw it.

"Draw another smaller square on top of it."

Give them time to finish their drawings. Don't answer any questions.

Ask some pupils to draw their suggestions on the board.

"Here is my drawing."

Show pupils your drawing.

"I did not give you any idea as to which way round to draw the squares or whether they were the same way up. To be able to copy my diagram you needed more information!"

"Let's try again. Draw a square with a horizontal base."

"Draw another square whose sides are one-half the length of the first square. The bottom side of this smaller square is part of the bottom side of the larger square."

Give pupils time to draw their diagrams and as they finish them select two or three that are significantly different.

"These diagrams are different so they can't all be right. The diagram I had in mind was like this:

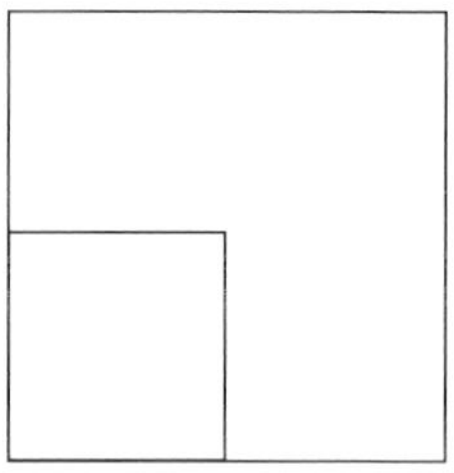

"I still did not give you enough information; I did not tell you where along the bottom line the smaller square was placed."

"Let's try again, and I'll try to be more precise. Draw a square with a horizontal base."

Give enough time for pupils to draw it.

"Draw another square whose sides are about one-third the length of the first square. The top side of the smaller square is part of the top side of the larger square and the top right-hand corner of both squares is the same."

Give pupils time to draw their diagrams. To any who are making obvious mistakes ask questions such as: 'Are the top right-hand corners the same?' 'Is the side of the smaller square one-third the size of the side of the larger square?'

Establish the correct answer.

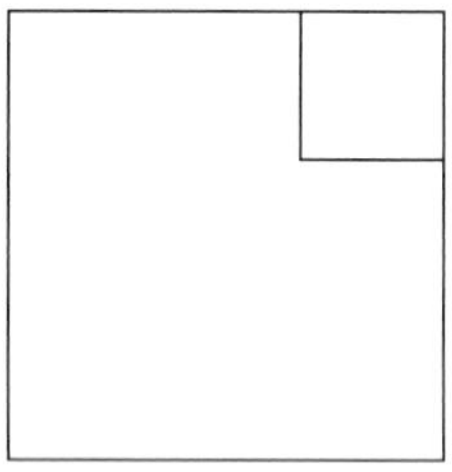

"This time I gave you enough information to draw the diagram accurately. We'll try one more before you have a go. Draw an equilateral triangle with the bottom side horizontal …"

"Draw a square below the triangle. The top side of the square is the bottom side of the triangle …"

"Draw straight lines from the top vertex of the triangle to each of the bottom vertices of the square …"

"I'll answer one question only if there is anything you are not sure of!"

Answer only *one question.*

When pupils have drawn their diagrams discuss any problems they have found.

Draw this diagram on the board.

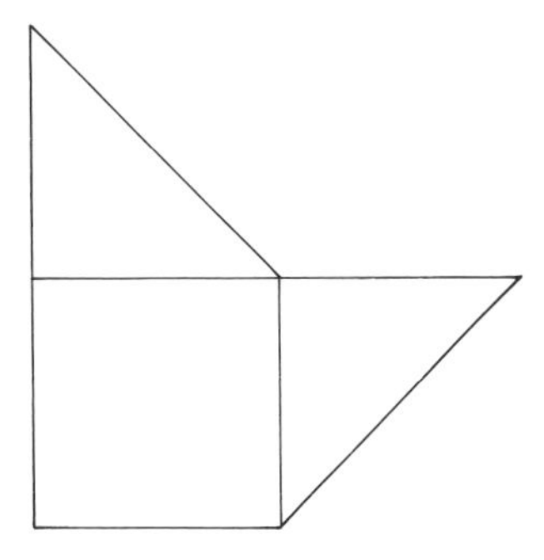

"How could we describe this precisely?"

Accept pupils' ideas and point out when they are incorrect or ambiguous by drawing alternative diagrams which fit their descriptions.

Bring out the fact that the diagram can be described correctly in more than one way.

"Now it is your turn. Sit back-to-back with a partner and one of you describe, as accurately as you can, the diagram on the card I give you. Discuss the diagram that has been drawn and then swap roles, with your partner describing a diagram from a different card. Discuss the diagram that has been drawn and then take turns describing a diagram that you have drawn. Start with a diagram with *only* two shapes and then try one with *only* three shapes."

Monitor the diagrams they draw and ensure that they are simple and are capable of being easily communicated to their partner.

Possible extension

- **Pupils could move on to more complicated diagrams that you give them, or diagrams of their own.**

Back-to-back

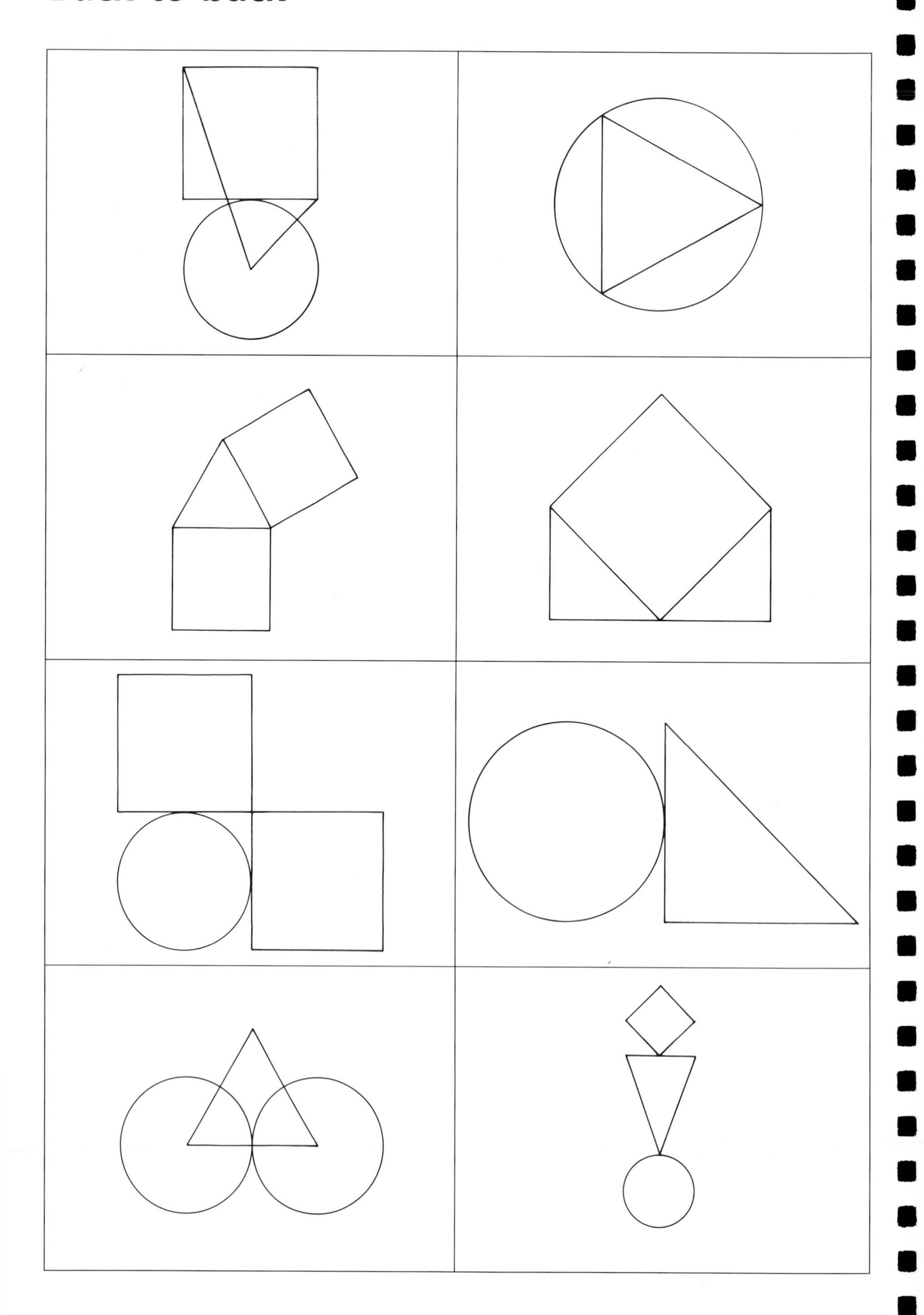

27 Points of view

Introduction The aim of this activity is to develop pupils' ability to 'see' hidden views of an object by visualising the object and imagining it turning or themselves moving round it.

Materials Cuboids such as multilink, chalk box, matchbox, wooden block, Oxo cube, Oxo box, LEGO, dominoes, etc.

Possible content Spatial orientation; views; plans and elevations.

Preamble The first time this activity is used the teacher should build the shape for a group of four to work with. Once pupils are familiar with the idea one of the group could decide on the shape for the others. This activity is best done by just one group of four at a time.

Before the pupils start this activity, build a shape from blocks, such as:

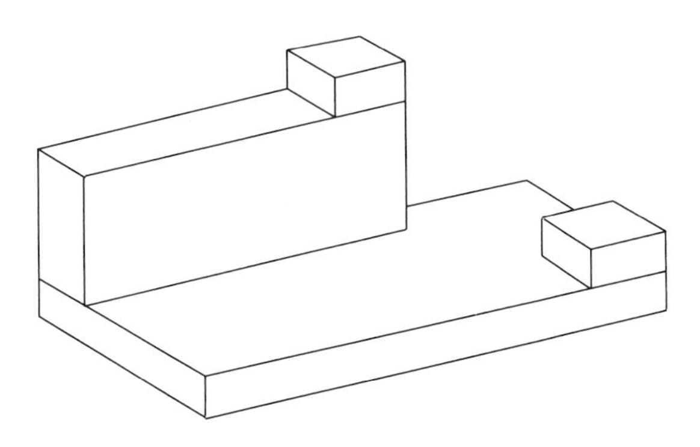

It should be on a table with access all the way round and straight edges parallel to the table edge. There should be a seat at each of the four edges. These should be labelled N, S, E and W.

"Walk round the shape and look at it from all four sides."

This may not be necessary for a high ability group.

"Sit down, one at each of the four sides.
Sketch the elevation you can see, not a three-dimensional diagram.
Label it N, S, E or W."

The term elevation may have to be explained. Allow enough time for pupils to draw their diagrams. Some groups may find square spotty paper useful.

"Without moving from your seat, try to sketch the elevation the person opposite can see. You may ask the person to describe it to you, but you must not go and look."

"When you have both finished compare your diagram with the one the person opposite you drew before."

Ideally the two sketches should be recognisably similar. If there are major differences, encourage the pupils to look from the other direction.

When all can draw the opposite elevation suggest they repeat the activity but try to draw the elevation seen by the person sitting on their right. If they have already compared all the drawings rearrange the model.

Possible extensions

- Make a model from given elevations.
- A group of pupils could make their own model, then draw the elevations. Their drawings could be a basis for another group doing the activity suggested above.
- Make a model which has two elevations the same. Draw all the elevations.

28 Visualising shapes

Introduction This activity can be dipped into. It involves pupils in visualising various shapes and answering questions about them.

Materials Various apparatus, as necessary, depending on the activity; possibly photocopies of sheets MI 16A and MI 16B.

Possible content Plane and three-dimensional shapes.

Preamble The first two activities are introductory and should be used with the whole class. The others can be used when appropriate. It is envisaged that *only a few will be done at any one time.* There are three types of activity. Those marked X are questions about shapes that pupils should be familiar with; those marked Y involve manipulations of shapes; and those marked Z involve moving shapes and where they are moving and why.

Apart from the introductory activities, pupils should be encouraged to discuss, in twos or threes, how they obtained their answers. The strategies for obtaining answers are very important and must be discussed. Pupils should be using arguments based on their mental images, and concrete apparatus should only be used as a last resort.

"Think of a capital letter X. How many straight lines do I need to draw it?"

Get a consensus that it needs two straight lines.

"Think of a capital letter P. What lines do you draw to get a capital P?"

Get an agreement that it is a straight line and a curve.

"I want you to think of all the capital letters and decide which you write using just three straight lines. Only write down these capital letters."

The purpose of this activity is to visualise the letters, decide whether they use three lines and then to write them down. No other writing should be allowed.

When most of the class have finished (about five minutes), discuss the capital letters in turn. Where there is disagreement, discuss whether it is because the letters are drawn differently.

"In your mind, picture a triangle."

Ask pupils to describe what they can see. Bring out the idea that different 'shapes' are possible and that the base does not have to be horizontal. Use the words scalene, isosceles, equilateral, acute-angled and obtuse-angled, if appropriate.

"Imagine a triangle again. Keep two corners fixed and move the other about. Imagine the different sorts of triangle you can get."

The purpose of this is to encourage pupils to realise that they need flexibility in their thinking. If pupils have difficulty, you could use the image of a triangle made from an elastic band (as in 'Elastic shapes').

The questions which follow can be dealt with in a variety of ways. The purpose of all of them is to encourage pupils to picture things in their minds' eye; the actual answers are of secondary importance. You will have to decide in what form you wish pupils to respond; whether they should give verbal replies or jot something down and how much discussion you want.

The material has been set out to enable you to use it in one of three ways – but you may think of others!

1 The teacher can 'read' a question to a group. Each member of the group then has to respond.

2 One pupil in the group reads the question and the others respond. All are involved in the subsequent discussion.

3 Individual pupils read the questions before joining together for group discussions. This reflects the fact that problems are often presented in a written form which requires some mental view before a response is produced.

The questions have been set in boxes so that they may be photocopied and cut up for use as suggested in 2 and 3 above.

For activities involving moving one shape around another, see *Using investigations*, page 64, 'Moving shapes'.

Visualising shapes

X Imagine a cube. How many edges does it have?	X Imagine a trapezium. How many obtuse angles does it have? Imagine other trapeziums. Do they all have the same number of obtuse angles?
X Imagine a noughts and crosses grid. How many lines do you need to draw it? Imagine a winning line of noughts. How many different winning lines are there?	X Imagine a cube. Concentrate on one corner. How many faces meet at this corner?
X Imagine an equilateral triangle. Picture its lines of symmetry. How many are there? Now try a square, rectangle, regular hexagon, …	X Imagine a parallelogram. How many acute angles does it have? Imagine other parallelograms. Do they all have the same number of acute angles?
Z Imagine a goat tethered to a post. What shape does it graze in the grass?	X Imagine a cuboid. How many faces does it have? How many edges does it have? Now try a tetrahedron, square-based pyramid, …
Y Imagine a square. What is the simplest way of cutting it into triangles? How many triangles do you have? What kind of triangles are they?	X Imagine a regular hexagon. How many pairs of parallel lines are there? Now try a square, regular octagon, regular decagon, …
Y Imagine two wooden cubes. Stick them together face-to-face. How many faces does the new shape have? How many edges does the new shape have? How many corners does the new shape have?	X Imagine a triangular prism. How many faces does it have? How many edges does it have? How many corners does it have? Now try a hexagonal prism, …
Z Imagine a bicycle being ridden along a flat road. Look at the hub of the wheel. What path does it trace out? Now imagine a point on the rim. What path does this trace out? You may like to consider the path of a pedal.	Y Imagine a triangle. Join another triangle on one of its sides. What different polygons can you make? You may vary the shapes of the triangles.

Visualising shapes

Y Imagine that you have a square piece of paper. Fold the corners to the centre. What shape do you get?	Y Imagine a rectangular piece of paper. Mark a point half-way along each edge and join them up. What shape do you get?
Y Imagine a triangle. Make one cut. What shapes do you get? What other shapes can you get by varying the triangle and changing the cut?	Y Imagine four straight lines drawn on a piece of paper. How many cross-over points are there? Move them about. How many cross-overs can you get? What is the minimum? What is the maximum?
Y We are going to build a framework tower with straws all the same length. Imagine a square made of four straws lying on the table. Put a straw vertically at each corner. Take these eight straws to be a unit. Put another one of these units on top, then another one. Finish it off with a square of straws on top. How many horizontal squares are there? How many horizontal straws are there? How many vertical straws are there?	X Imagine a cube. Join two corners through the centre of the cube. How many lines are there like that? Join the centre of two faces through the centre of the cube. How many lines are there like that? Join the mid-points of two edges through the centre of the cube. How many lines are there like that?
Z Imagine two dots in space. What is the locus of points equidistant from these?	Y Imagine three cubes stuck together to make an L shape. How many faces does this new shape have? How many edges does it have? How many corners does it have? Now try this with four cubes making a T, seven cubes making an H, …
Y Imagine a cube. Cut a piece off with a straight cut. What shaped cut face have you got? Try different cuts. What shaped faces can you get?	Y Imagine a cube. In what ways can you cut this into two identical pieces with a straight cut? Which of these ways are planes of symmetry?
Z Imagine a fixed square. Imagine another square that can move about which has a red dot somewhere in it. Slide the moving square so that it is just touching the fixed square edge-to-edge. The moving square moves all the way around the fixed square without tilting or turning. What line does the red dot trace out?	Z Imagine a cube. Imagine a line joining the centre of the top face to the centre of the bottom face. Now make the cube turn around this line. What is the order of rotational symmetry? How many axes of symmetry are there like this?

29 Patterned polyhedra

Introduction The aim of this activity is to develop pupils' ability to visualise how a net will fold to make a polyhedron.

Materials Sheets MI 17, MI 18 and MI 19; coloured pens or pencils; scissors; glue.

Possible content Two-dimensional representation of three-dimensional objects; mathematical language concerned with shape.

Preamble This activity can be used with a whole class, a small group or individuals. Making the shapes is of secondary importance; the emphasis should be on visualising what will happen when the net is folded. It is helpful for discussion with pupils if you have a net of a cube already cut out.

Give each pupil a copy of sheet MI 17. Pupils should not have access to scissors at this stage.

"Later you will be making a cube from net A.
Imagine what it will look like when you have finished.
Think about which edges will touch when it is folded."

Allow a few minutes for pupils to imagine the net being folded to make a cube.

"When you have a clear picture of which edges join together, colour or label the pattern so that the parts which are going to join are the same."

Stress that they should think carefully before they start colouring or labelling.

It may be appropriate to talk about faces, edges and vertices or about adjacent edges. Allow plenty of time for pupils to think about the pattern and colour the parts of the pattern which will touch.

The finished cube could be used as a basis for discussion about parallel and perpendicular edges and faces.

The finished cubes can be mounted as a three-dimensional wall display.

This activity could be repeated using the other nets at other times.

Patterned polyhedra 1

Cube

Imagine these nets folded to make cubes.
Colour or mark the patterns so that those sections which will join up are the same.

Cut out the net.
Make sharp creases along the fold lines.
Fold the net into a cube, but do not glue it.
Check whether it has matched up as you expected.

If it has not matched up, see where the colours are wrong, then open it out again to see why.
With your net flat, mark the sections you think will match. Fold to check.
Glue the flaps and make the cube.

A

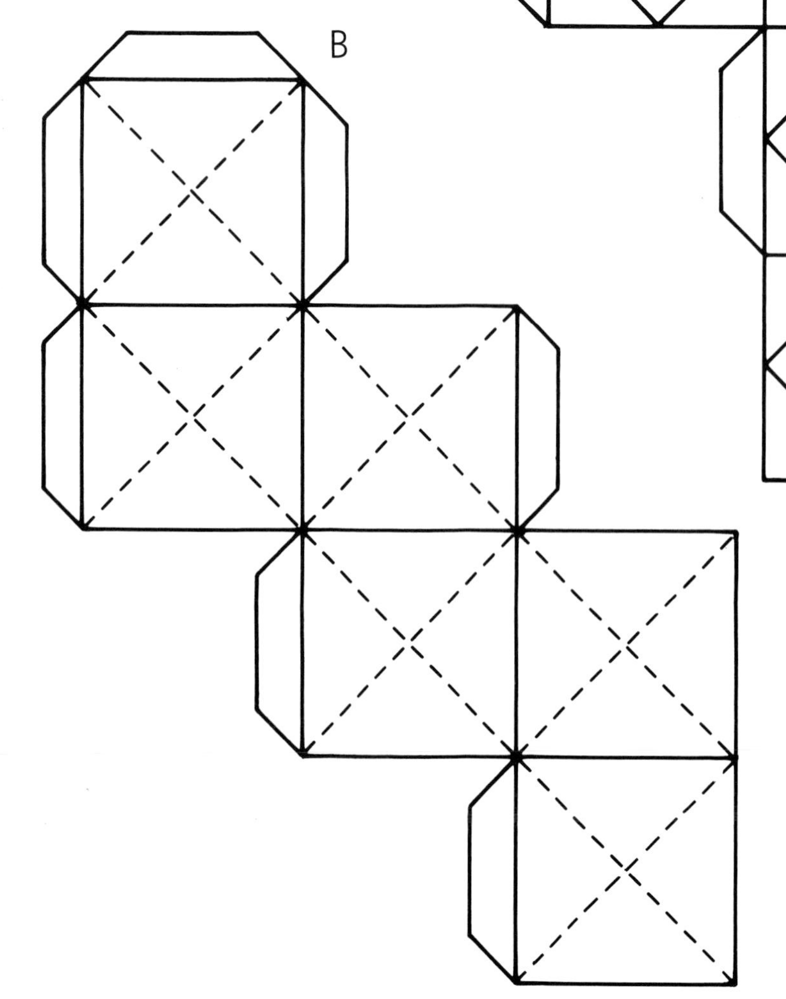

Patterned polyhedra 2

Dodecahedron

Imagine this net folded to make a dodecahedron.

Colour or mark the snakes so that those which will join are the same. Then cut out the net and make the dodecahedron.

Patterned polyhedra 3

Cuboctahedron

Imagine this net folded to make a cuboctahedron.

Colour or mark the patterns so that those which will join are the same. Then cut out the net and make the cuboctahedron.

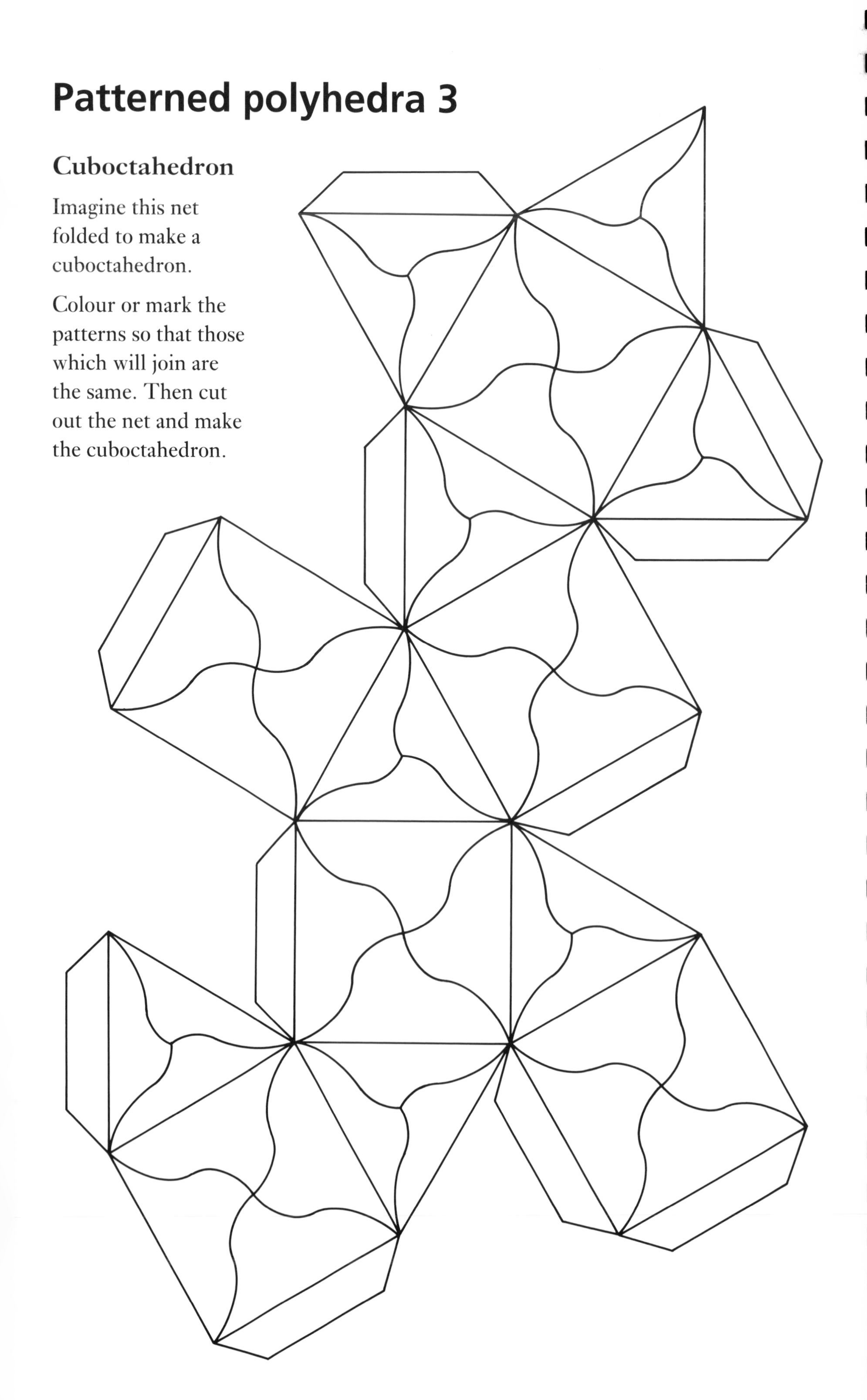

30 Shadows

Introduction This activity is designed to help pupils with visualisation and manipulation of shapes in three dimensions. It may be regarded as a fairly quick, informal activity, although it could be expanded.

Materials A way of casting shadows, such as OHP, slide projector or light box; some shapes cut out from stiff card, for example square, scalene triangle, scalene quadrilateral.

Possible content Language of plane shapes. It may be possible to illustrate the process of making and disproving conjectures.

Preamble This activity is suitable for a whole class or group. However, this depends on the 'shadow-making apparatus' available.

Use the square card to cast some shadows.

"By moving this square around I can make differently-shaped shadows."

Make three or four different shapes. Allow sufficient time for pupils to grasp that the shadow is not always the same shape as the object.

Put down the square.

"What shaped shadows can I make?"

Discuss suggestions.

"Do you think it is possible to make a triangular shadow with a square?"

Accept answers from the group; encourage discussion to develop. Ask a pupil to explain why it is not possible, using the square if necessary.

"What sorts of triangular shadows can you cast from a scalene triangle?
For example, could you cast a triangular shadow which is isosceles?"

This activity could also be used for introducing vocabulary associated with triangles.

Reach a consensus as to what triangles are possible.
If appropriate, allow pupils to investigate the problem practically.

Possible extensions

These can be attempted by the whole class or small groups.
It is intended that the questions should be answered mentally, but some practical checking may be necessary to resolve differences.

- What shadow shapes can be made from a hexagon?
- What shadows can be made from an equilateral triangle?
- Choose a shape which has at least one pair of parallel sides.
  Will the shadow always have the same number of parallel sides?
- Imagine a non-square shadow cast by a square. If you made a copy of this shadow onto card, could you use it to cast a square shadow?
- What shadow shapes can be made from: (a) a cube, (b) a triangular prism, (c) a sphere?

31 Folding paper

Introduction Pupils are encouraged to imagine what shapes will be made by different cuts across folded paper.

Materials Worksheet MI20; scissors; A6 paper rectangles, enough for at least five per person. (They could be cut from scrap paper.) Do not hand out any of these at the start.

Possible content Symmetry and angle properties of polygons.

Preamble Pupils may wish initially to cut the paper without thinking first, but after a few cuts they usually realise that this leads to repetitions.

Hold up a piece of paper, A4 or larger. Fold it in half, and half again, as shown below.

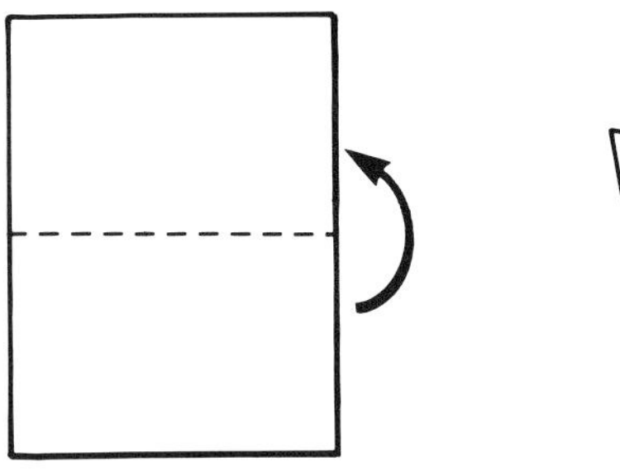

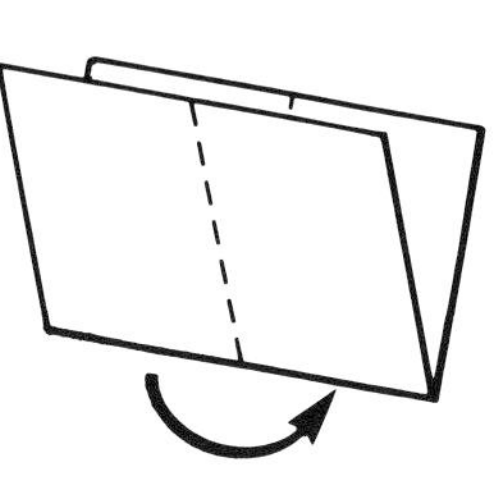

Hold the scissors and paper to show clearly that you are going to cut from one folded edge to the other.

"I am going to cut straight across this folded corner.
What shapes do you think I could make when I open it out?"

For each shape suggested, ask exactly how the cut should be made. Make the cut, then open out the shape. Discuss the results.

The relationship between squares and rhombuses may arise here. If appropriate, use the creases in the paper to discuss the property of diagonals bisecting each other at right-angles.

Discuss whether it is possible to get any shape other than a rhombus.

"What if I make *two* cuts like this?"

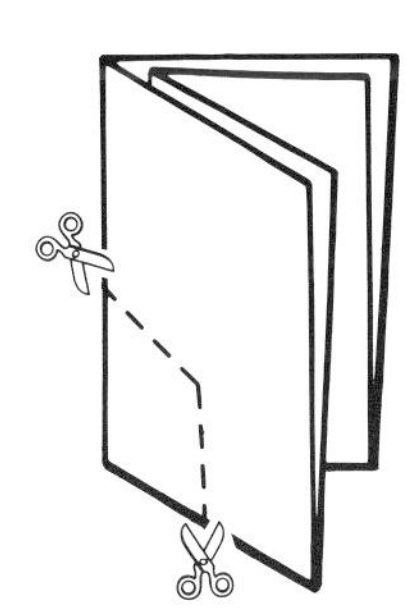

Now demonstrate that there will be one cut from each folded edge.

"Imagine the ways I could make two cuts to cut off the folded corner. Imagine the shapes I could make."

"You will be working in pairs to make different shapes. Each time, think where you want to cut and visualise the shape."

"Before you make your cut, sketch the cut and the shape you expect to get, using the worksheet. Then make the cut and see if you made the shape you expected."

"Try to make a square, an octagon, a hexagon and a star using two cuts for each shape. You will only have *five* pieces of paper, so think before you cut."

Give out the worksheet, scissors and five pieces of paper to each pair.

"Keep all your different shapes – you may be asked what you got, and how."

Allow time for all the pairs to try and make the shapes. Restricting the number of pieces of paper usually eliminates random cutting, but more can be given if really necessary.

For each of the shapes in turn, ask someone to describe the cuts they made, and to hold up the shape that resulted. Ask if anyone else made the same shape a different way.

This may lead to a discussion about regular and irregular shapes and their angles. The octagon either is regular or has alternate pairs of sides the same length.

"How does the square ... hexagon ... star relate to the octagon?"

Pupils may make comments such as 'To get the square you cut at right-angles so it is a straight line when you open it out', 'A hexagon is an octagon with two angles that are straight lines' or 'Each shape must have two lines of symmetry'.

You may wish to ask questions such as 'Are they all octagons?'

"Did anyone get any other shapes by accident, apart from those discussed already? Don't say what the shape was."

If this is the case, then ask for a description of the cuts. Ask if anyone can visualise what shape this would give. Check that this is what really happened.

With some groups it may be appropriate to finish the activity here. A homework task could be to record all the shapes possible with explanations of why the cuts gave those shapes.

"What happens if the second fold is not at right-angles to the first?"

Demonstrate how to fold the paper.

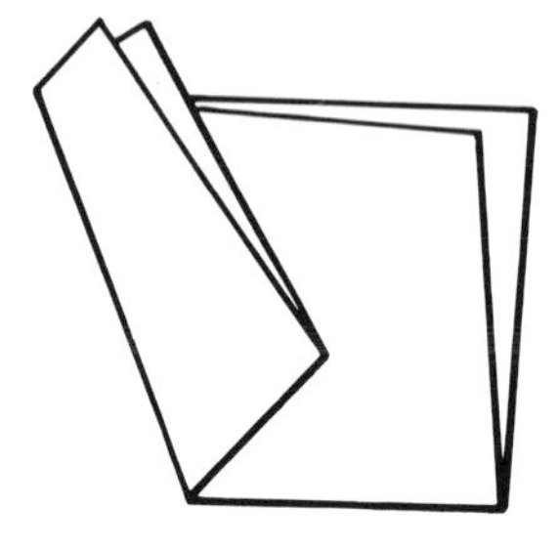

"In pairs, try one cut and see what you can get; then try two cuts. Think about the shape you will get before you cut or unfold the paper."

"Think about the different types of shapes you might make, and what cuts you would need to make them."

Cuts made perpendicular or parallel to the various folds produce triangles, pentagons, etc., which can lead to a discussion of angle and symmetry properties of shapes.

Some pupils develop a system or record systematically (for example all those cases where one cut is perpendicular to a fold). This leads to finding some unexpected results, such as a seven-sided shape.

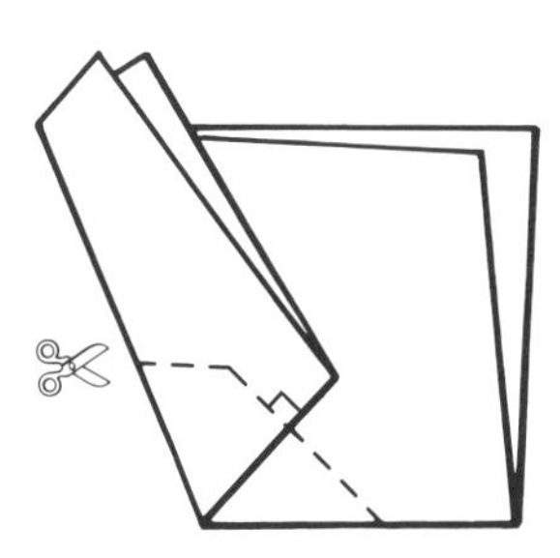

At the end of the activity the different folds and cuts can be classified. The cut-outs, with their folds marked, can be mounted on coloured paper to make an interesting wall display.

Name ....

MI 20

Folding paper

Before making each cut, show where the cut will be and sketch the shape you predict you will make.

Cut	Predicted shape	Were you right?
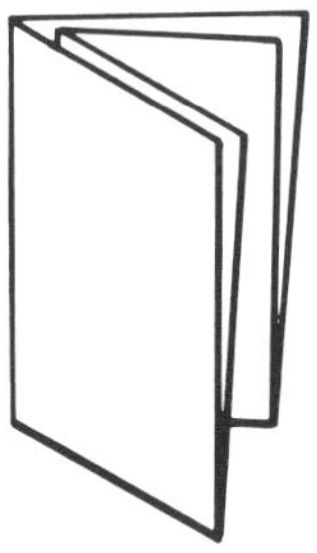		

32 Turnaround

Introduction This activity is designed to help pupils manipulate geometric images.

Materials A 5 × 5 pinboard and elastic band or equivalent; worksheet MI21, to be handed out in advance.

Possible content Rotation in three dimensions.

Preamble This activity is for a whole class. For recording, pupils could work individually or in pairs. If they work in pairs, which will generate more discussion, then hand out one worksheet per pair.

On your 5 × 5 pinboard form a right-angled triangle as shown.

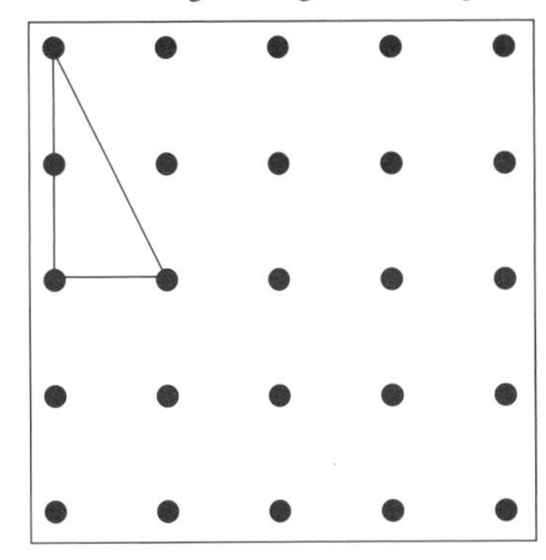

Hold the pinboard facing the class.

"What can you see?"

Accept all contributions.

"I am going to turn the board round half a turn. Imagine what you will see."

Allow pupils a few seconds.

"Tell me what you will be able to see."

Accept all contributions without comment.

Turn the pinboard through half a turn.

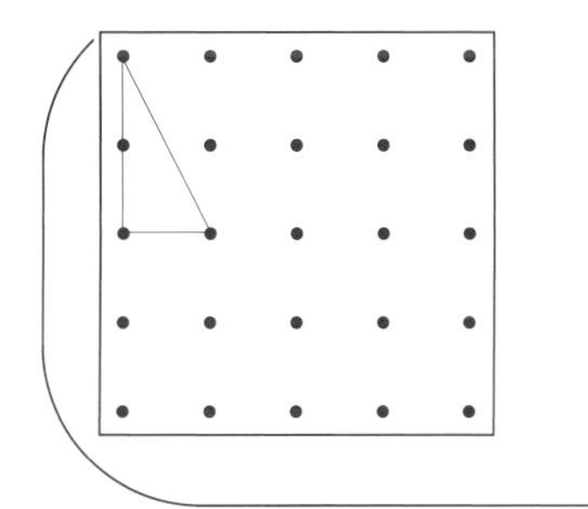

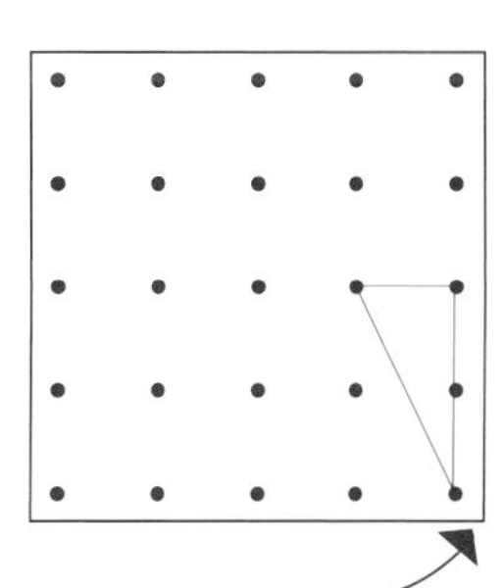

Now discuss pupils' previous contributions as appropriate.

Start again with the pinboard in the original position and the pins facing the class.

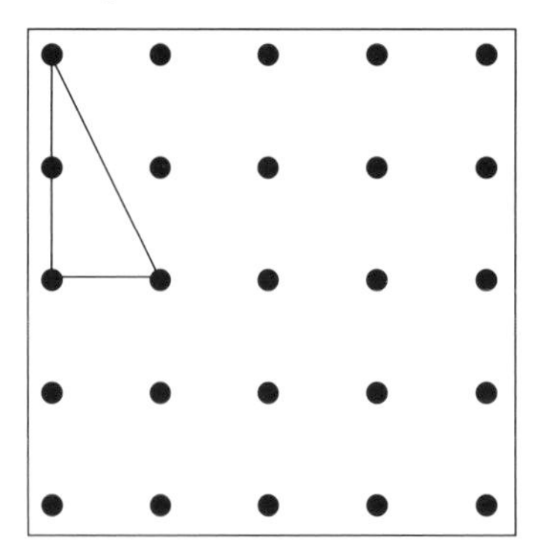

Give the board a half-turn about a horizontal axis through the centre. The pins should now be facing you.

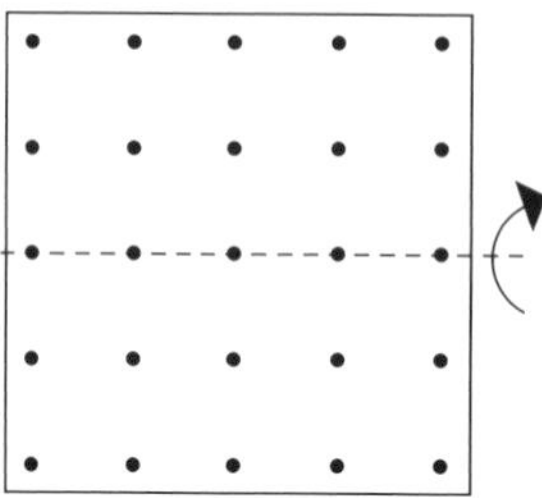

"Imagine what you think I can see."

Allow about 30 seconds.

"Record in one of the squares on your worksheet what you think I am looking at. In other words, where is the triangle from my point of view?"

Allow time for recording.

"Tell me what you have drawn."

Accept all contributions, encouraging pupils to be as precise as possible. Ask other pupils if they agree or disagree, and why.

Turn round (your back to the class) to show them what you can see on your pinboard.

Start again, with the pins facing the class and the triangle in the top left-hand corner.

Give the board a half-turn about a vertical axis through the centre so that the pins are facing you.

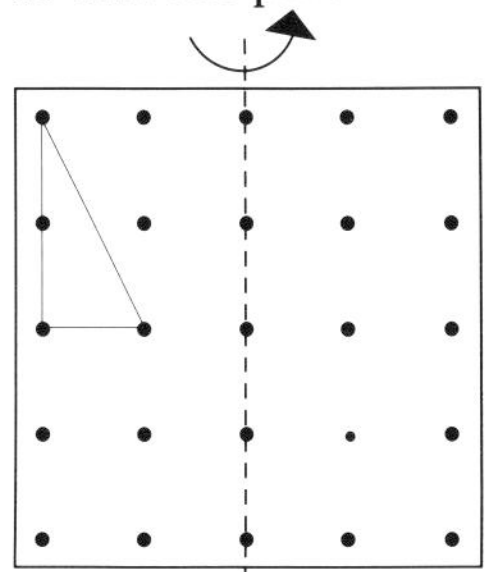

"Imagine what you think I can see."

Allow 30 seconds.

"Draw on the next blank square where you think the triangle is from my point of view."

Allow time for recording.

"Tell me what you have drawn."

Accept and discuss all contributions.

Turn round (your back to the class) and show them what you can see.

Repeat the above procedure, only this time give the board a half-turn about one diagonal.

Repeat, this time turning about the other diagonal.

Possible extensions

- Start with the triangle in a different position, for example:

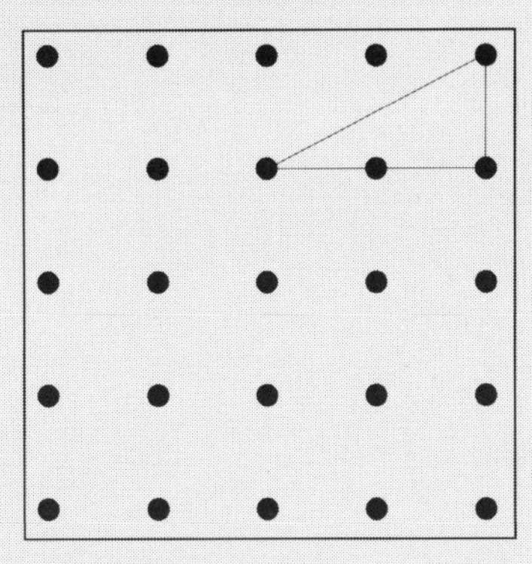

  Repeat the same operations as used previously.

- Use a combination of turns. For example, give the board a half-turn about the horizontal (pins now facing you). Give the board a half-turn (pins facing you all the time) and then ask pupils to describe what you can see.
- It would be fun at the end to have a test! Present pupils with a few situations. They record their ideas on the blank squares (worksheet). Remember ... make a note of the turns and the answers as you go along. It is quite easy to forget which way you turned the board and in what order!

Name ....

Turnaround

33 Elastic shapes

Introduction Pupils are asked to close their eyes and imagine the vertices of shapes moving to form other shapes.

Materials About one metre of elastic tied in a loop.

Possible content Classification and properties of two-dimensional shapes.

Preamble All the instructions should be read very slowly, to allow plenty of time for pupils to imagine the movement.

"Close your eyes.
Imagine an elastic band round three pins, forming an equilateral triangle with a horizontal base."

"Imagine the pin at the top of the triangle moving sideways until it is vertically above one of the pins at the base. Think about the shape of the new triangle. Now move the top pin horizontally the other way until it is above the other base pin. Think about the new triangle."

"Fix the pictures of the three triangles in your head, then open your eyes."

"What can you tell me about the two triangles you made by moving the pin?"

Collect suggestions on the board for discussion. These are likely to include 'They have right-angles' and 'They are reflections'. If there is disagreement, or if anyone cannot 'see' the triangles, demonstrate the movement using the elastic and three pupils holding pencils as pins.

"How else could you move the top pin?"

Collect suggestions and decide which to use – probably 'It could move up and down.'

"Close your eyes. Imagine your original equilateral triangle.
Make the top pin move up slowly, then down to where it started."

"Now move the top pin down slowly until it touches the base of the triangle, then slowly back up again. Open your eyes. What can you tell me about what you saw?"

Again, collect comments. It should be possible to discuss properties of equilateral and isosceles triangles.

Depending on the group, ask them to imagine the top pin moving in some of the other ways suggested, for example along one of the sides of the triangle towards one of the other pins, or to imagine one of the base pins moving.

When appropriate, perhaps on a different occasion, move on to four pins, as in the following activity.

"Close your eyes. Imagine you have four pins in a horizontal row. There is an elastic band round the row. Think about where you should move the inner two pins so that the elastic band makes a square. Fix the square in your mind, then open your eyes."

"What does your square look like?"

There are two possible squares.

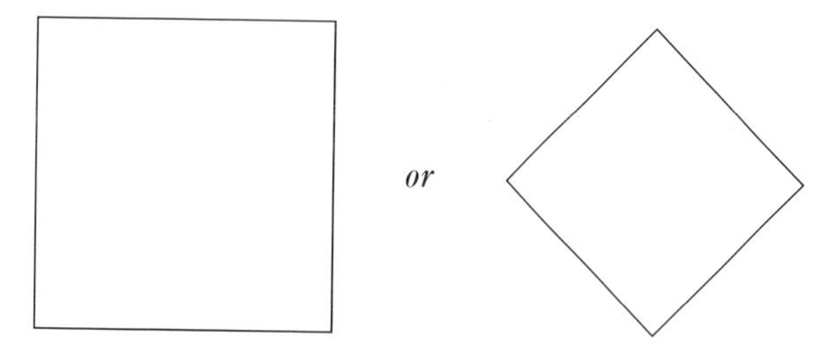

Ask pupils to describe and demonstrate the moves for both.

The pupils could now work in twos or threes. Ask them to visualise which pins they should move, and how, to make a kite, a rhombus, a parallelogram, a rectangle and a trapezium. Taking turns, they should describe a move or series of moves to their partners, who should say what shape is made.

When everyone has had a go in their group, a few could be asked to try their moves on the whole class.

Properties of quadrilaterals could now be discussed – both symmetry properties and the relationships between different quadrilaterals.

34 Walkabout

Introduction The aim is to develop pupils' ability to visualise orally-given routes and to enable them to give clear and precise directions without the use of a map.

Materials Possibly plans of the school and local maps or street plans.

Possible content Use of left and right; possibly the points of the compass.

Preamble This is a teacher-led activity for a group or the whole class, followed by work in small groups. The first sets of directions should be kept short and straightforward to ensure that all pupils can follow the instructions and are clear about the types of directions to give. The routes chosen will depend upon the geography of the school and those given below are merely intended as examples to give an idea of the suggested complexity. Think through the routes you are going to use in advance, but do not use a script. It would be useful to tell pupils a week before the activity that they will be doing some work about routes around the school.

"Close your eyes.
In your mind, follow the directions I am going to give you and decide where you will end up. Go out of the classroom, turn right … go along the corridor, through the double doors … stop at the first door on the left … what door is this?"

Keep the first set of instructions very simple so that there is little disagreement about the destination.

"Close your eyes again … follow this route. Go out of the classroom, turn left … go along the corridor … down the stairs … turn left … through the double doors … stop at the first door on the right … what door is this?"

This route should be a little longer than the first, but still simple enough to ensure that the group can be successful. A short discussion may be necessary.

Now try a more complicated route.
Use directions like those given above.

Avoid giving hints like 'past the library'.

Discuss the route and destination and ensure that everyone agrees about the outcome.

"Close your eyes again.
Go out of the classroom, turn left … go along the corridor. You start to go down the stairs but you find that the route is blocked by fire and smoke. … Work out a safe route to the nearest exit."

Give pupils time to plan their route and then discuss alternative suggestions.

"How many other classes would have to leave by this exit if there was a fire on that staircase?"

Questions like this can broaden the scope of the activity.

"Now work in pairs."

"Each of you plan a route from this classroom to somewhere in the school. Keep your route fairly simple.
Write down your route.
Take it in turns to read out your route to your partner while they close their eyes and work out your chosen destination."

Give them plenty of time for this. It is important to check that their written routes are clear and not too complicated.

"Now repeat this, but this time planning a route from the school entrance to the staff room, sports hall, here ..."

Starting in a different place may make the activity a little more difficult.
Any routes given as homework should be kept simple.
Emphasise that the route should be planned mentally before it is written down.

Possible extensions

- Plan a route from the school entrance to the town centre, supermarket, library, etc.
- Plan sensible fire exits for different parts of the school.
  These could be checked against the 'official exits'.

35 Worms

Introduction This activity is designed to help pupils with visualisation in three dimensions.

Materials Sheets MI22 and MI23. OHP transparencies of these might be useful for large group work. Sufficient multilink for about 20 cubes per pair of pupils.

Possible content Working with two-dimensional representations of three-dimensional objects.

Preamble This activity is suitable for a whole class or group of more able pupils. The first part should be seen as an optional warm-up activity giving practice in counting cubes where the divisions between them are not distinct.

You should make the shapes shown on sheets MI22 and MI23 before the start of the lesson.

Show the whole group sheet MI22.

"Here are drawings of some cubes.
How many cubes are there in shape A?"

Allow sufficient time for all pupils to obtain an answer.
Discuss how they arrived at their answers and what strategies they employed.

Repeat with the other shapes.
This activity should allow you to decide whether the main activity is too difficult for some pupils.

Arrange the pupils in pairs and give them the multilink cubes and sheet MI23.

"Just by looking, can you tell me the least number of cubes you need to connect these two end-faces on drawing A?"

Indicate the two shaded faces on drawing A.
Collect answers and establish the correct one (four).

"Working in pairs, I want you to make the connecting shape or 'worm' from multilink."

Check to make sure that all pairs have attempted this.

"Now check to see if your worm fits my model."

Allow pupils to see if their worms connect up correctly with your own previously made model.

Throughout the activity, you may find it useful to allow testing of a worm only after the pupils have made it.
Do not allow trial and adjustment at the actual model.

Depending on the level of success, repeat this activity with another drawing.

"In your pairs, repeat the activities for the rest of the shapes shown on the sheet."

"Test your worms against my models."

If there are sufficient cubes, pupils could make the models themselves. Pupils in each pair could then take it in turns to be 'worm maker' or 'model maker'.

Possible extensions

- Pupils could make worms which physically connect to the model at both ends, taking into account the projections on the cubes.
- Pupils could be asked to make several different worms for a particular model. This activity could be used as an informal introduction to three-dimensional vectors.

Cubes

A

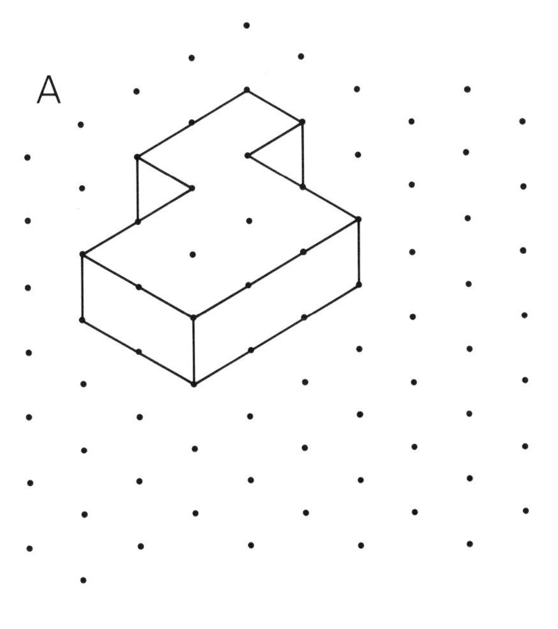

B

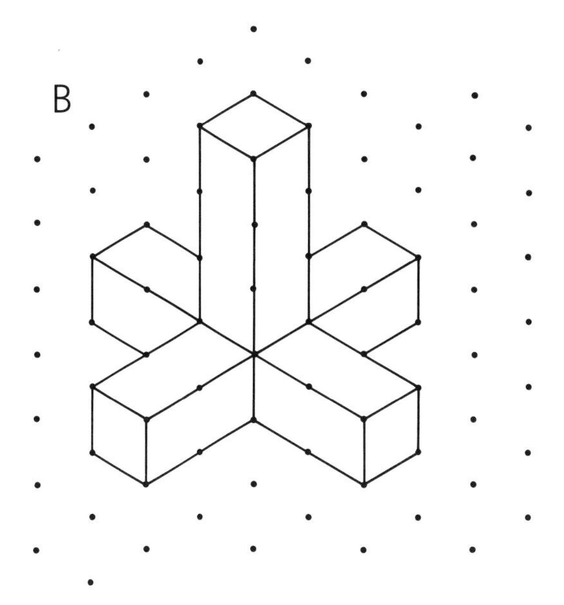

C

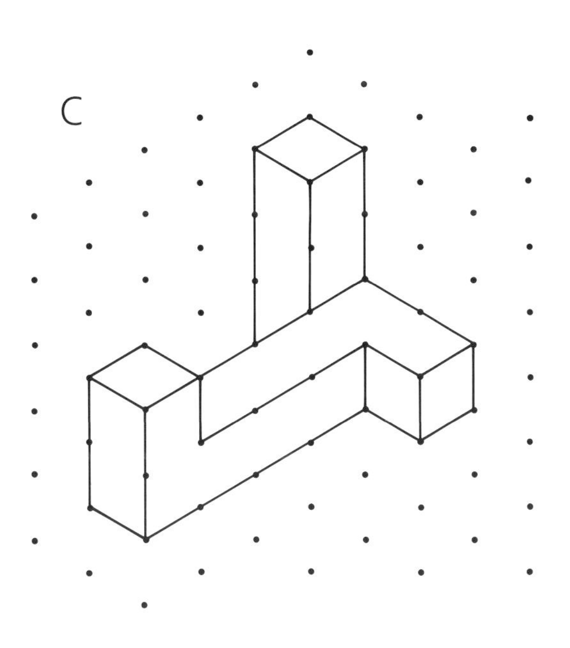

D

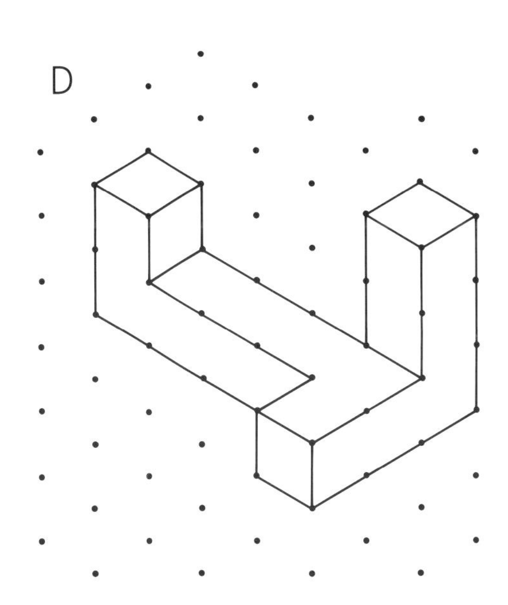

MI 23

Worms

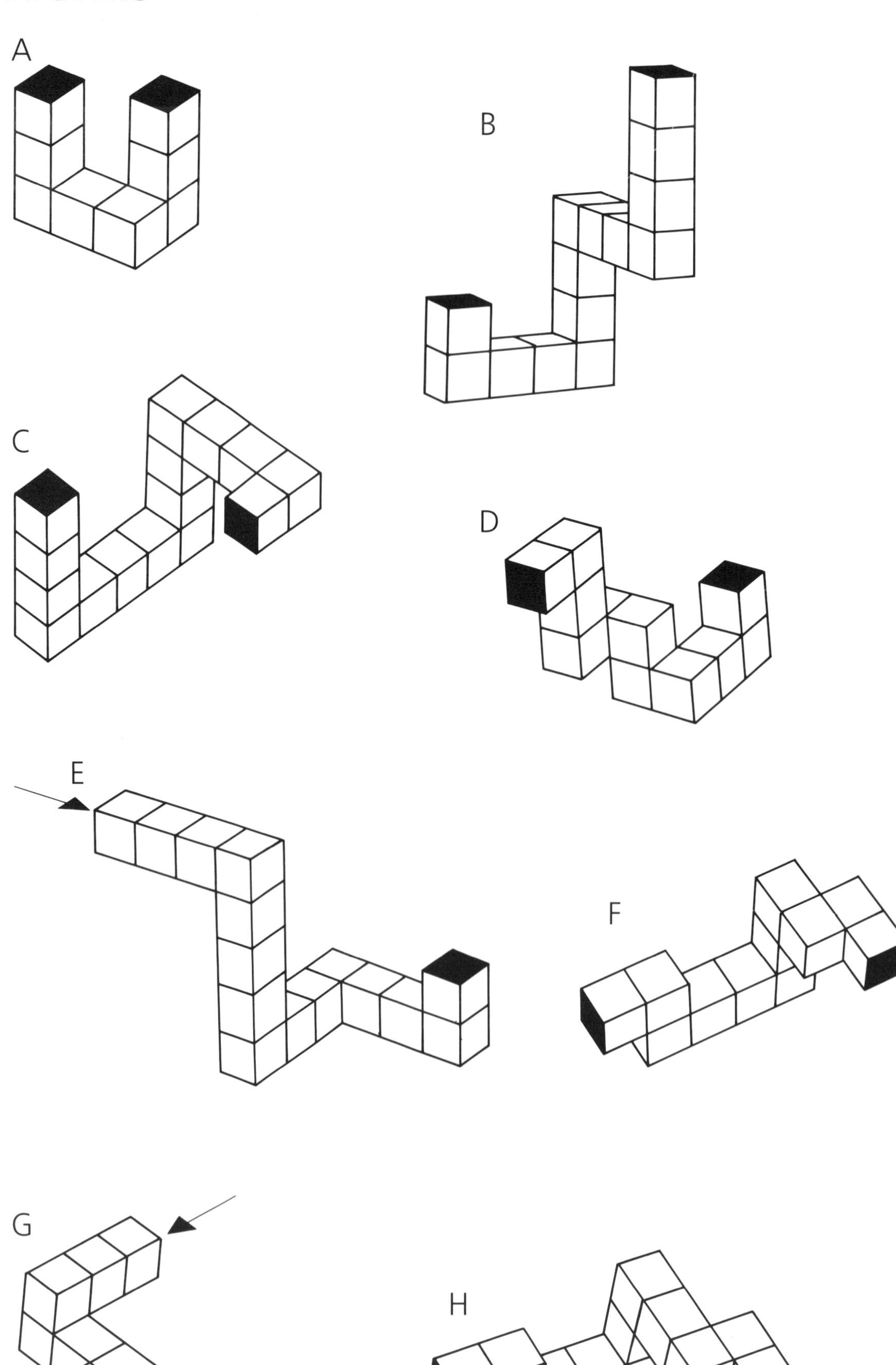

36 Rings and strings

Introduction These activities are designed to help pupils with visualisation of spatial movement. 'Rings' may be seen as an alternative to 'Strings', depending on ability. 'Rings' is easier than 'Strings'.

Materials Sheets MI24 and MI25.

Preamble These activities are suitable for a whole class.

Show the group sheet MI 24, indicating box 1.

"Look carefully at these three rings.
Which, if any, of the other rings are attached to ring A?"

Collect and discuss answers without using concrete materials.

Repeat the activity with the other sets of rings on the sheet.

"How can we tell if any rings are attached to a particular ring?"

Collect any ideas on the board. Pupils will probably home in on the 'unders' and 'overs', but do not force this.

"An interesting configuration is the last set of rings. Are these three rings attached to each other?"

Split into small groups.
Allow each group to investigate any rules which may have arisen previously, such as number of cross-overs, patterns in 'unders' and 'overs'. Allow groups to test their rules by drawing their own arrangement of rings, as well as using those on the sheet.

Should any pupils experience difficulty with this, they could use string. Colouring the rings may also be useful.

Show the group sheet MI 25.

"Look carefully at the string shown in box 1."

"If you pulled the two ends would you make a knot?"

Discuss the answers without using string.

Repeat the activity with the other strings on the sheet.

"How can we tell whether a drawing is of a knot or not?"

Collect suggestions on the board. Continue as for the 'Rings' activity.

Possible extensions

- Which drawings are of the same knot?
- Draw pictures which look different but give the same knot.
- Look up knots which have an everyday use.

Rings

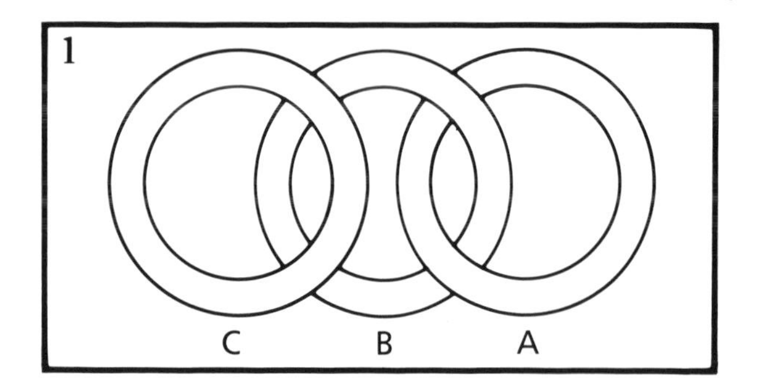

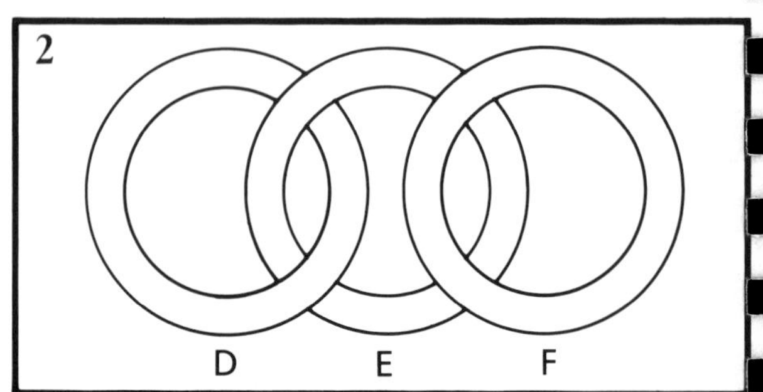

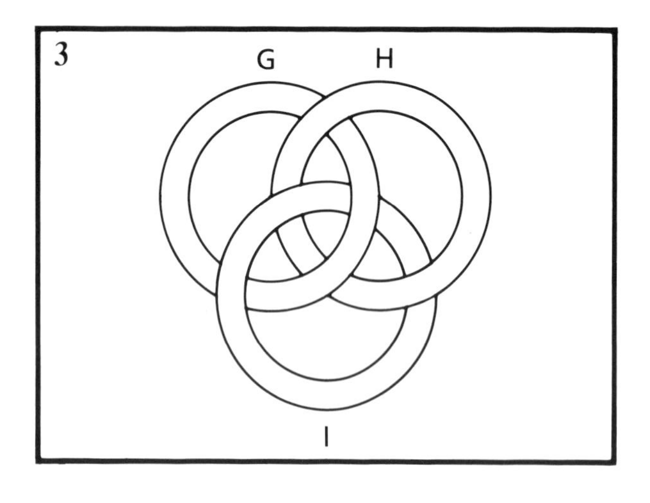

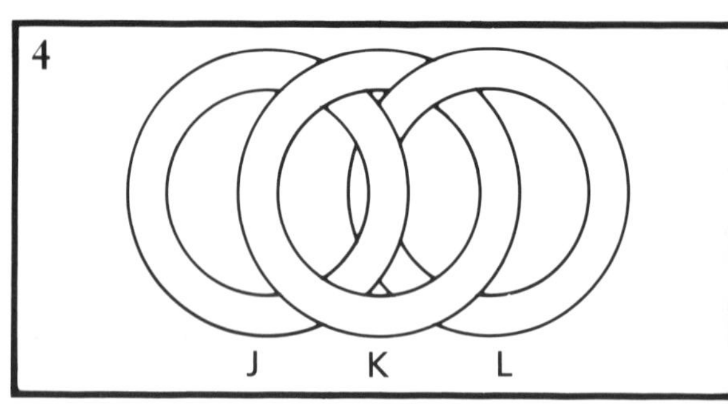

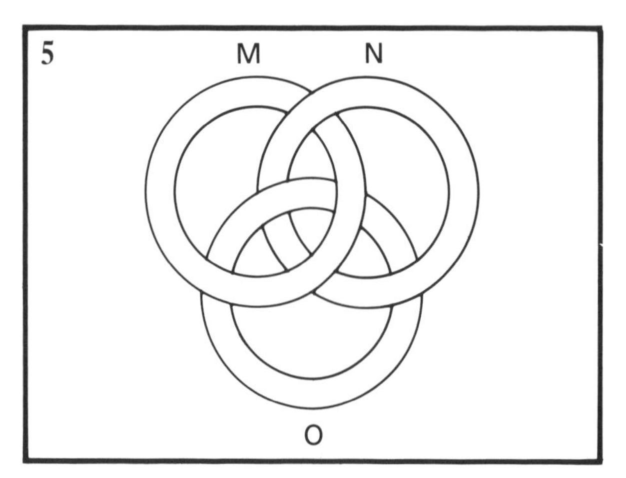

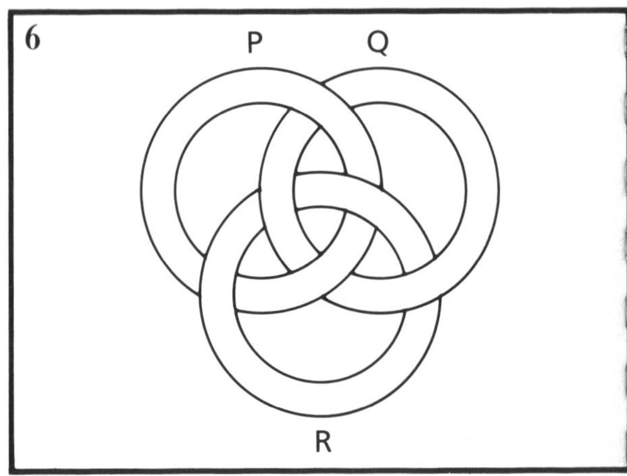

Strings

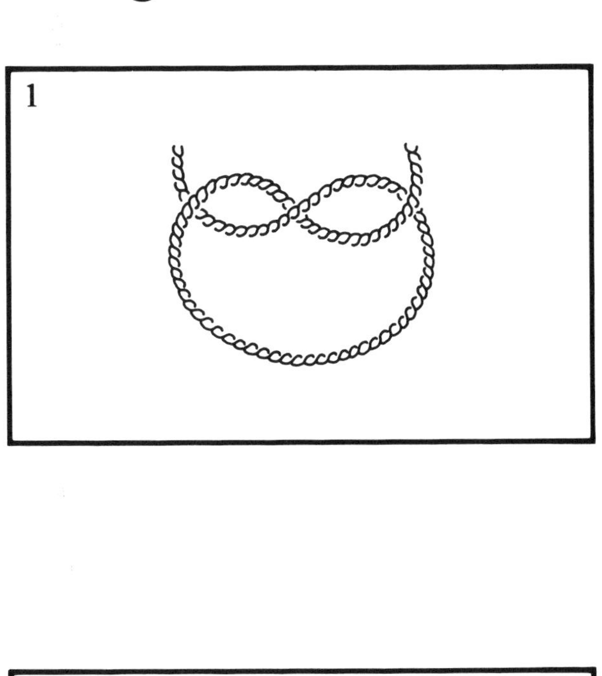

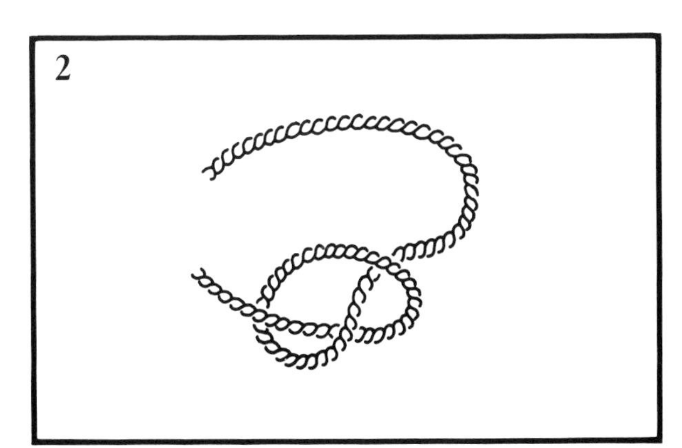

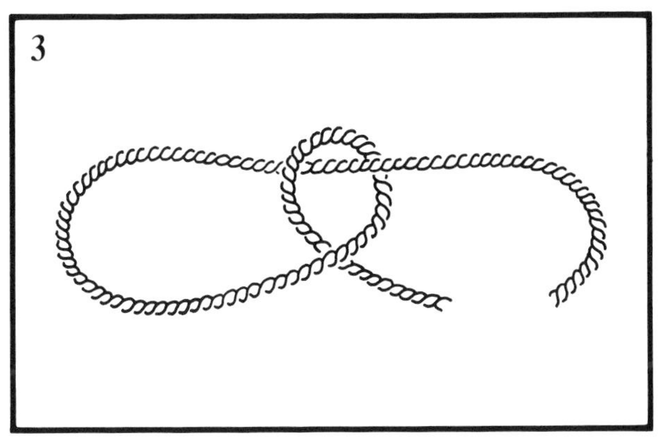

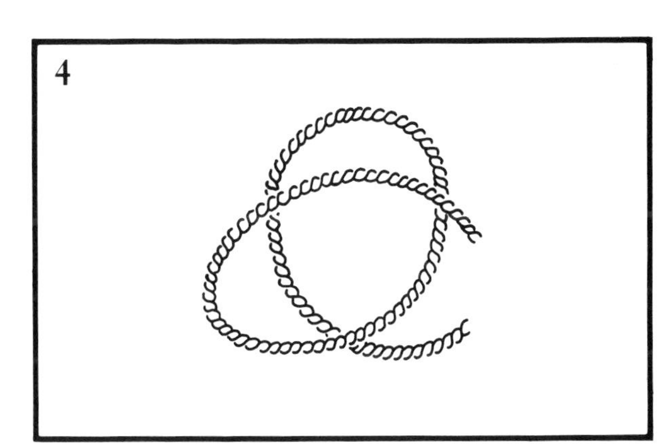

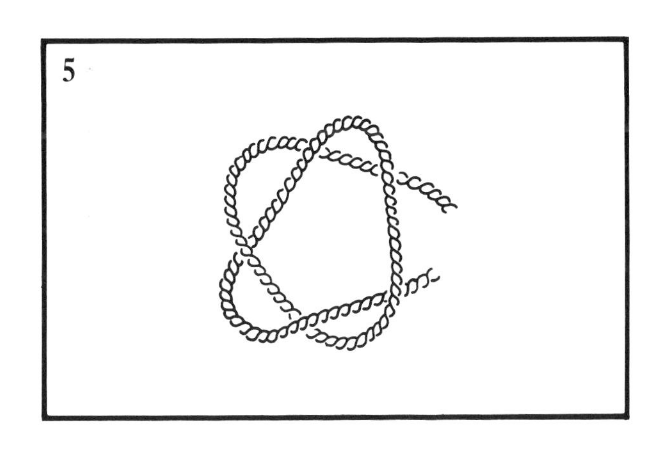

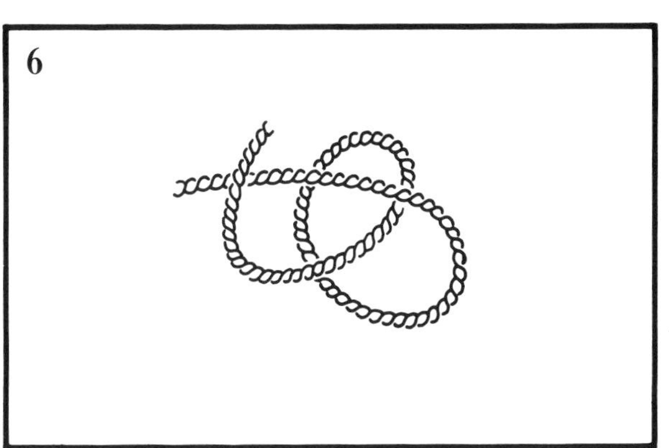

37 Square dance

Introduction This activity is designed to help pupils develop spatial awareness and memory skills.

Materials Possibly, a square board (of side about 30 cm) and four 'Post-it' notes labelled 1, 2, 3 and 4; worksheet MI 26 for extension work.

Possible content Rotation, reflection.

Preamble This activity can be carried out entirely in the mind. However, for some groups the use of the board and 'Post-it' notes may be helpful. Possible ways of using these are given at the end of the activity.

"Close your eyes.
Imagine a square … put the number 4 in the top left-hand corner … and 2 in the top right-hand corner … now put 1 in the bottom left-hand corner … and 3 in the bottom right-hand corner."

Pupils' mental image should be:

4	2
1	3

You might like to check the image by getting a pupil to draw it on the board. Erase it before proceeding.

"I am going to ask you to move these numbers by giving you some instructions."

Start off with a single instruction, such as:

"Swap top left with top right."

"Swap top right with bottom left."

"Move all numbers one corner clockwise."

"Swap the top row with the bottom row."

"Reflect in a line from top left to bottom right."

Each time ask pupils to tell you where each number is. Establish the correct answers.

"Now put the numbers back in their original positions."

You may need to remind pupils what these were.
Repeat the activity with other single instructions.

When all pupils are confident about following one instruction, then use two or three (or more) instructions of the kind given above. It is advisable to have written these down in advance, so that you can repeat them if necessary when discussing the solutions.

Remember to 'reset' the numbers before each set of instructions.

Possible uses of the board and 'Post-it' notes

The board and notes provide a concrete example of the mental image. The 'Post-it' notes allow easy interchange of the numbers. They can be used at any point where you feel such concrete reinforcement is necessary, but remember that this is intended to be a *mental* exercise. Some possible uses are:

- as a preliminary demonstration before doing any mental work;
- as a working model when you give out instructions (make sure pupils cannot see what you are doing) so that you can show the final answer;
- to show intermediate stages when discussing solutions.

Possible extensions

- How many different arrangements of the four numbers are possible?
- Present a starting situation and a finishing one and ask what would be the smallest number of instructions required to change from one to the other,

  for example

4	2
1	3

  to

1	4
3	2

  Worksheet MI26 provides eight examples. They could be done in pairs, or as homework.

- These two extensions could be combined to find minimum moves from

  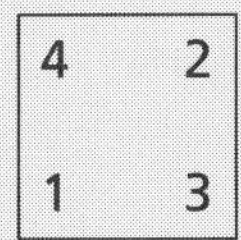

  to each of the other possible arrangements.

Name ....

MI 26 Square dance

For each pair of squares, write down how to change the first to the second in the smallest number of moves.

1	2		1	4
4	3		2	3

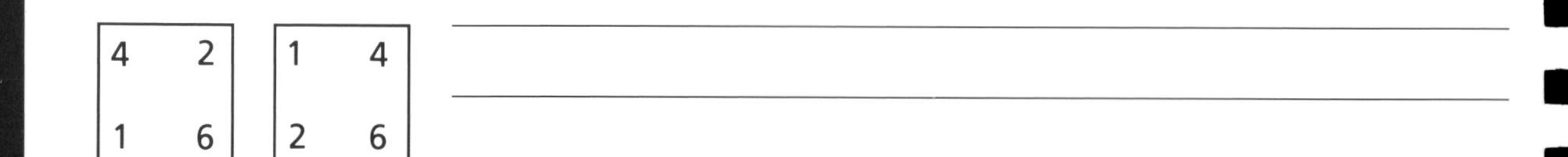

p	q		p	s
s	r		r	q

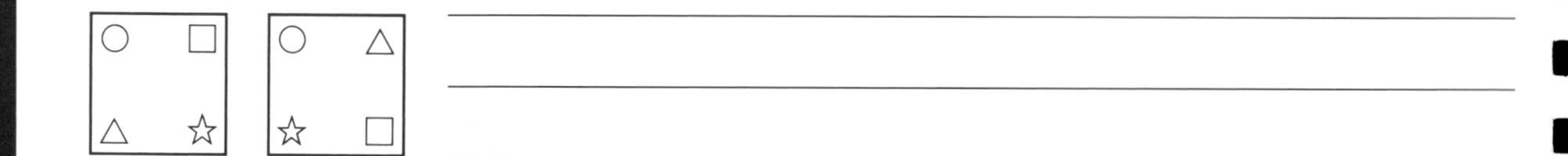

R	B		M	R
M	D		D	B

T	o		y	o
n	y		n	T

IV	III		I	II
II	I		IV	III

π	θ		θ	α
α	λ		π	λ

38 Sorting out Eth Porter

Introduction This activity is a first step towards the construction of logical proofs or explanations. It aims to overcome pupils' difficulties in setting out clear reasoned solutions.

Materials Sheets MI 27 and MI 28, scissors.

Preamble The activities given here should be seen as starting points for your own suggestions. You could create activities of this type whenever you do a suitable piece of work. Frequent practice is important to build up pupils' confidence. This activity is best done in pairs, to encourage discussion.

Give each pair of pupils a copy of sheet MI27. Get them to cut out the 'shape' statements cards. The cards should be shuffled and placed face up on the desk, so all are visible. Tell pupils to arrange the cards to make a sensible paragraph. Encourage them to discuss their solution. Ask if there is more than one possible paragraph that makes sense.

They could be asked to produce a similar set of cards to try out on another pair of pupils. Do not let them make their paragraphs too long or complicated. They should compare the solution produced by the other pair with their own. If it is not the same, they should discuss whether both versions are valid.

Repeat this activity with the 'counters' statements cards.

Now hand out worksheet MI28 to each pair, explaining that some pupils did a survey and wrote a report. The statements of their report have been jumbled up and listed on the worksheet. Ask each pair to put the statements into a sensible order by writing down a list of statement numbers. As the statements cannot be moved around physically, this is not an easy task. Once they are satisfied that they have a complete survey report, they could write it out to check that it makes sense and that they have not missed out any of the statements. Ask if it is possible to omit any of the sentences without affecting the meaning of the report. If so, which ones are they?

Ask each pair to produce its own jumbled report, for another pair to sort out. Make sure that it is not too long.

You could ask those pupils who have some understanding of algebra to put the steps of an algebraic solution of an equation in order. For example:

Solve $3x - 14 = x + 8$.

$2x = 8 + 14$ (1)

$x = 11$ (2)

$3x - x - 14 = 8$ (3)

$2x = 22$ (4)

$2x - 14 = 8$ (5)

Do this type of activity whenever the opportunity arises.

'Shape' statements cards

Squares and rectangles are shapes.	but
Rectangles also have	Triangles do not.
they do not have sides.	They have
four sides.	three sides.
Circles are also shapes,	Squares have four sides.

'Counters' statements cards

in a bag.	the red one.	I have three counters
the yellow counter.	Paul gets	One is red,
yellow.	left	another is blue
Alison pulls out	in the bag?	and the last one is
What colour counter is		

Name ....

The report

Without writing the sentences out, put these statements together to make a sensible report on the survey. Just write down the statement numbers.

1 We decided to include five colours.
2 It was a fine day and Ms Macintosh was in a good mood.
3 We talked about what to do and decided to do a survey.
4 Mike made the tally.
5 It rained when we did the tally.
6 I called out the colour of a car.
7 We did not take any notice of bikes or buses.
8 It would be interesting to try the experiment at the same time on another day.
9 The sun came out at three o'clock.
10 We had to find which colour of car was the most popular outside the school.
11 This is a summary of our results.
12 It was market day so most of the drivers may have come from out of town.
13 Our results show that red was the most popular car colour.
14 We drew a bar chart.
15 1 green
16 13 black
17 Mike went outside first to see what colours to include.
18 We did the tally from two o'clock to three o'clock in the afternoon.
19 The two of us, Janet and Mike, did this work.
20 We ignored any lorries.
21 In the tally chart we labelled the colour columns red, black, white, blue and green.
22 13 blue
23 Perhaps the car drivers at this time were not a fair sample.
24 We started this investigation on Tuesday.
25 Ms Macintosh was very pleased with our results.
26 at least on the day and at the time we did the survey.
27 67 red
28 2 white

Write the order of the statements here:

....

....

....

39 Words and their meanings

Introduction The poster is used as a stimulus for discussion.

Materials Poster of mathematical words.

Possible content Any area of mathematics.

Place the poster so that all the pupils can see it. Make no comment, simply ask the pupils to tell you what they think the poster is about *without* referring to or naming any particular item on the sheet. When you think they have a feeling for what it is about ...

Ask them to choose, silently and in their minds, one of the objects on the sheet and to think about how the drawing expresses the meaning of its related word or shape and how they might explain it to another person.

Ask them in turn to say which drawing they have chosen and to give their explanation. Make no comment, except to clear up any misunderstanding that might arise.

Brainstorm a list of mathematical words and record contributions on the board. Add those on the poster to the list because pupils may think up different ways of expressing them.

Ask the pupils to use no more than ten of the listed words and to make a similar poster for display.

40 Pictures with errors

Introduction This activity is designed to encourage discussion and the use of mathematical language.

Materials Poster of a street scene with errors.

Possible content Many areas of the mathematics curriculum.

Preamble The poster can only be examined by a few pupils at a time. Groups could be sent up while the rest of the class is doing other work.

Ask each group to examine the poster and to agree a list of errors. They should make a written list between them. Tell them that the list should say which objects have something wrong with them, to explain what is wrong and, if possible, how it could be put right. There will probably be plenty of discussion but the main difficulty is likely to be that the ideas are not written down coherently.

Allow each group as long as they need. When all the groups have finished, ask for an error from each group in turn and ask the other groups to comment. The group describing the error also has to say how to correct it. Each group must identify an error which has not already been mentioned and the turns should continue until all the suggestions have been discussed. There is likely to be a natural competition between the groups about who has found the most errors.

Possible extension

- Ask pupils to draw their own pictures with errors – a picture of a kitchen would give a variety of possibilities. These could make an interesting wall display.

Bibliography

SMP materials where spatial visualisation is used

SMP 11–16 booklets

Maps, plans and grids	Left and right are used to give directions.
Bearings and journeys	Pupils draw a journey described by a series of bearings and distances.
Views	Pupils draw and interpret views of models.
Mathematics frompictures	Views and aerial photographs are compared with maps.

SMP 11–16 textbooks

G5	chapter 3 Photographs and maps of the same places are compared.
G6	chapter 7 Pupils interpret maps to follow and give directions.
B2	chapter 1 Views of a building are interpreted in relation to a plan.
Y2	chapter 10 Pupils visualise intersections of lines and planes represented in two dimensions.

Using investigations

Moving shapes	Pupils are asked to visualise the track mapped out by objects moving according to a selection of rules.

New stretchers

1a Match point	Isometric drawings of models made from cubes and wedges are used to decide which pictures represent the same model.
1b Houses	Visualisations of houses from diagrams are used to decide which pictures are of the same house.
6 Cube routes	The possible routes along edges from one vertex of a cube to another are considered.
8 Views and bases	This involves building three-dimensional models from cubes, given the plan and two elevations.
2 Nettles 9 Networks 18 Face nets	These consider how faces, edges and vertices join when nets are folded to make cubes.
15a Cube puzzler 15b Tetra puzzler	Different orientations of a lattice cube and tetrahedron are compared.

Using groupwork

Tower game	A group of pupils build a copy of a shape made from cubes. Only one of the group is allowed to see the original shape and that person is not allowed to touch the cubes, only to describe the shape to the rest of the group.
Pyramids galore	After an initial discussion pupils visualise and design the nets needed to make unusual shapes such as prisms with irregular cross-sections or skew pyramids.

Other references

Association of Teachers of Mathematics, *Readings in mathematical education*, ATM 1988 (ISBN 0 900095 72 5)

The Mathematical Association, *Mental methods in mathematics*, MA 1992 (ISBN 0 906588 27 8)

The Open University, *Working mathematically on mental imagery with third formers* (Study pack code PM647C)